Here Be There Dragons

ALSO BY RICHARD A. LOVETT

Fiction

Phantom Sense & Other Stories (with Mark Niemann Ross)

Million Dollar Marathon (with Philip Maffetone)

Nonfiction

Alberto Salazar's Guide to Road Racing (with Alberto Salazar)

Alberto Salazar's Guide To Running (with Alberto Salazar)

The Essential Cross-Country Skier (with Paul Petersen)

The Essential Touring Cyclist

Freewheelin': A Solo Journey Across America

Here Be There Dragons

Exploring the Fringes of Human Knowledge, from the Rings of Saturn to the Mysteries of Memory

Richard A. Lovett

Strange Wolf Press
Portland, Oregon USA

Strange Wolf Press
Publishers of eclectic books since 2012
Portland, Oregon

ISBN: 978-1540443434

Front cover art: NASA
Front cover layout: Richard A. Lovett
Back cover layout: Richard A. Lovett
Interior design: Richard A. Lovett
Proofreading and editorial: Holly Hight
Author photo: Vera Jagendorf

CONTENTS

INTRODUCTION
&
ACKNOWLEDGMENTS

When I was in fourth or fifth grade, my parents gave me a book. It wasn't the first book they'd ever given me, but it might be the one that most strongly shaped who I became. Or maybe it just resonated with who I was already becoming; in that type of formative memories, it can sometimes be difficult to separate cause from effect.

It was called *All About Famous Scientific Expeditions* and was a collection of stories about not just science, but the adventure of pushing scientific exploration to the fringes of the possible. Enthralled, I joined naturalist William Beebe (the Depression-Era counterpart to Stephen Jay Gould) on the first descent into the ocean abyss, as he plunged 3,000 feet into the depths in a claustrophobic diving bell called a bathysphere. I joined archaeologist Howard Carter as he caught the first glimpses of King Tut's tomb; scaled Alaska's Mt. Denali with Hudson Stuck (whose own book, *Ascent of Denali* would later become an adult favorite); dashed for the North Pole with Robert Perry; and searched for Komodo dragons in Indonesia. Each story gripped me in its own way, but collectively what they did was to instill in me a desire to know what lies beyond the next horizon …whether that horizon be physical or intellectual. By the time I was in middle school, I was sure I was going to be a scientist.

At the same time, I was devouring classic science fiction: Asimov, Clarke, Heinlein, Kornbluth, Bradbury, Verne, Wells, and pretty much anything else I could get my hands on. If there was one thing I wanted more than to be a scientist, it was to be a science fiction writer.

Life subsequently took some unexpected turns, and while I got a degree in astrophysics, I then went to law school…and collected a Ph.D. in economics. Obviously, I didn't really know what I wanted to do…so I wound up in the perfect field for generalists: I became a free-lance journalist.

In my early years I dabbled in many fields, but what paid the rent were trade-magazine articles about such dull-sounding subjects as regulatory toxicology, nutritional epidemiology, biotechnology, microbiology, analytical chemistry, and food safety. Then an assignment took me to a meeting of the American Association for the Advancement of Science. During a break in the food-safety programming, I randomly stuck my nose in another session…and found the latest high-resolution pictures of Mars. Behind another door were images from ocean depths beyond anything Beebe had dreamed of. It was like rediscovering my *All About* book…except that instead of reading about history, I was watching current events. In the process, I realized that somehow, I had come full circle. I might not have become a scientist, but I had become a science writer.

A few years later, my middle school dream of writing science fiction also came true, as I sold the first of what are now fifty-two stories, most of them to *Analog Science Fiction and Fact*, the magazine in which many of the stories I'd loved as a child first appeared. At the same time, I started selling science articles to *Analog* (contributing to the "fact" part of its name). Intriguingly, as of the date I'm writing this introduction, there are also fifty-two of those.

Fiction writers compile their work into short-story collections. This book is the equivalent, in science writing: eighteen of my personal favorites, set on the fringes of known science. They're a diverse lot, covering everything from space exploration to archaeology. There's even one about wood-peckers.

"All About" science, it is not; if you want that, read Bill Bryson's *A Short History of Nearly Everything*. This is a sampler—a format that allows me to go into more depth in each story than would otherwise be possible. Think of it as eighteen stories about eighteen cool topics picked by a science fiction writer. Though, if you're not into science fiction, don't worry; the primary focus here is the science.

As for thanks, there are too many people to mention. Many of these stories grew out of symposia at science meetings that graciously provide reporters with free admission. They also would never have existed without the hundreds of scientists who shared their wisdom, either at meetings or on the phone.

Other science writers helped me learn the ropes. Particularly strong thanks go to Victoria Bruce (for dragging me to my first American Geophysical Union meeting), Harvey Leifert, Christina Reed, Stanley Schmidt, Emily Hockaday, Trevor Quachri, and the numerous editors I've worked with at *New Scientist, Nature, Science, Cosmos, and National Geographic*. Special thanks also go to Holly Hight for eagle-eyed proofreading and editing. Any errors that might now exist are ones introduced after she looked at it, as I struggled to force Microsoft Word to do things it didn't particularly want to do, such as wrap text around the cool art supplied by the scientists and by NASA. Additional thanks go to Liana Bernard and Krystin Hawkins for putting up with endless geeky conversations, and to the long-ago teachers who nurtured my interest in science and exploration. Thanks also to Mark Niemann-Ross, my partner in crime at Strange Wolf Press, who first convinced me that yes, we really could produce small-press/indie books. So far we've published work from four authors (often in collaborations), with the primary goal of producing fun, eclectic reading. More are on the table.

Finally, I want to acknowledge the original sources of the articles that became chapters in this book. Specifically:

"Planets (Oops, Planetoids) X, Y, Z and W: What the Kuiper Belt Teaches About the Dawn of the Solar System," *Analog Science Fiction and Fact*, April 2012.

"Forensic Seismology: The Big Science of Minor Shakeups," *Analog Science Fiction and Fact*, April 2004.

"Here Be There Dragons: The Ivory-Billed Woodpecker and Other Mysteries of an Explored Planet," *Analog Science Fiction and Fact*, October 2008 (under a slightly different title).

"Dance of the Rings," *Cosmos 39*, June/July 2011.

"The Day the Sun Exploded," *Analog Science Fiction and Fact*, November 2012.

"Geology, Geohistory, and "Psychohistory": The (Continuing) Debate Between Uniformitarians and Catastrophists," *Analog Science Fact and Fiction*, May 2009.

"From Atlantis to Canoe-Eating Trees: Geomythology Comes of Age," *Analog Science Fiction and Fact*, September 2009.

"Dividing the Invisible: The History of *Atomos*," *Cosmos Magazine* (online), 8 November 2012 and 15 November 2012 (as "History of the Atom").

The Prehistory of Global Climate Change," *Analog Science Fiction and Fact*, March 2005.

"From Fimbulwinter to Dante's Hell: The Strange Saga of Snowball Earth," *Analog Science Fiction and Fact*, Jan/Feb 2006.

"Plate Tectonics, Goldilocks, and the Late Heavy Bombardment: Why Earth isn't Mars or Venus," *Analog Science Fiction and Fact*, December 2009 (under a slightly different title).

"Poisons, Temperature, and Climate Change: Will Global Warming Make Everything Else Worse?" *Analog Science Fiction and Fact*, December 2011.

"Energy 2063: The Future of Global Energy Production," *Cosmos 50*, May/June 2013.

"The Search for the World's First Equestrians," *Analog Science Fiction and Fact*, November 2007.

Shake, Rattle, and Roll: Is Missouri Really America's Most Dangerous Earthquake Zone?" *Analog Science Fiction and Fact*, September 2011.

"Paleolakes, Jøkulhlaups, and Mobergs: What Iceland Reveals about 'Wet Mars,'" *Analog Science Fiction and Fact*, February 2003.

"The Transience of Memory: We Really *Can* Remember It for You Wholesale," *Analog Science Fiction and Fact*, June 2004.

"Saturn's 'Jet-Propelled Moon and the Search for Extra-Terrestrial Life," *Analog Science Fiction and Fact*, September 2014.

All of these articles, however, have been updated to include advances since the time of their initial publication.

PLANETOIDS
X, Y, X, AND W

What the Kuiper Belt teaches about the dawn of the Solar System

When Pluto was dethroned as a planet, many thought it silly. Where once there had been nine planets and a bunch of asteroids, there were now eight, the asteroids . . . and Pluto. What was the point? But the fact is that beyond the orbit of Neptune lies a shadowy zone of frozen worldlets orbiting so slowly that few have made complete circuits since America ceased to be a collection of British colonies. This region, called the Kuiper Belt, has fascinated science fiction writers since its existence was first suggested fifty years ago. Ironically, however, the person it's named for, astronomer Gerard Kuiper, apparently didn't actually believe it existed. Yes, he proposed an ancient belt of planetoids in this region as the source of the icy moons of Jupiter, Saturn, Uranus, and Neptune—but in the ensuing billions of years, he presumed, gravitational perturbations had caused most if not all of its denizens either to fall inward and be captured as moons or be completely ejected from the distant Sun's feeble hold.

But it turns out that Pluto (and its large moon, Charon) aren't the only survivors. At least a thousand such bodies are now known, and the tally is growing. Several are large enough that any might have been dubbed Planet X, had astronomers stumbled across them decades ago in their then-vain search for a tenth planet. From a science-fictional point of view such planetoids are intriguing because they're all considerably bigger than any asteroid but Ceres, offering a whole new array of worlds for exploration. But they're also intriguing to scientists because, like the asteroids, they're remnants of the early Solar System—leftovers from the planet-forming era, now frozen in time. The largest is Eris, named for the Greek goddess of strife and discord. It orbits three times farther out than Pluto and, at 2,300 kilometers in diameter, is close enough to Pluto's size that it has indeed produced (mild) discord…over whether it is or isn't bigger than Pluto. (The latest estimates make Pluto monarch of the Kuiper Belt, but not by much.) Behind Eris, the list includes:

- Haumea (shaped like an American football, 2,000 kilometers long). It's named for the matron goddess of the island of Hawaii, in honor of Mauna Kea Observatory, where it was discovered;

16

- Sedna (another football, 1,600 kilometers long), named for the Inuit goddess of the sea, who dwelt below the frigid Arctic waves;
- Makemake (a 1,500-km sphere), named for the fertility god of Easter Island (because it was discovered shortly after Easter);
- Pluto's moon, Charon (1,200 km);
- Orcus (950 km), named for an Etruscan/Roman god of the dead;
- Quaoar (890 km), named for the creator god of the Tongva (or San Gabriel) people, native to the Los Angles area, where it was discovered; and
- Ixion (650 km), named for an outlaw god of Greek mythology.

Finding more is a priority for several astronomers, including Matthew Holman, an astrophysicist at the Smithsonian Astrophysical Observatory at Harvard University, whose team works with a telescope called Pan STARRS-1, atop a 10,000-foot peak on Maui, Hawaii. With an aperture of only 1.8 meters, Pan STARRS-1 isn't huge. What makes it unique is that it has an extremely wide field of view, capable of examining in a single image a swath of sky forty times larger than the full moon. It also has four 1.4 billion-pixel digital cameras, the largest ever built as of 2014. All this technology was designed to scan for potential Earth-impacting asteroids, but it's also ideal for spotting Kuiper Belt objects down to 300 to 500 kilometers across. (Bigger telescopes can see much smaller ones, but only by focusing on much narrower regions of the sky.) In its first months of test runs in 2010, Pan STARRS-1 found ten new Kuiper Belt objects. Holman reported these findings at a January 2011 meeting of the American Astronomical Society in Seattle,

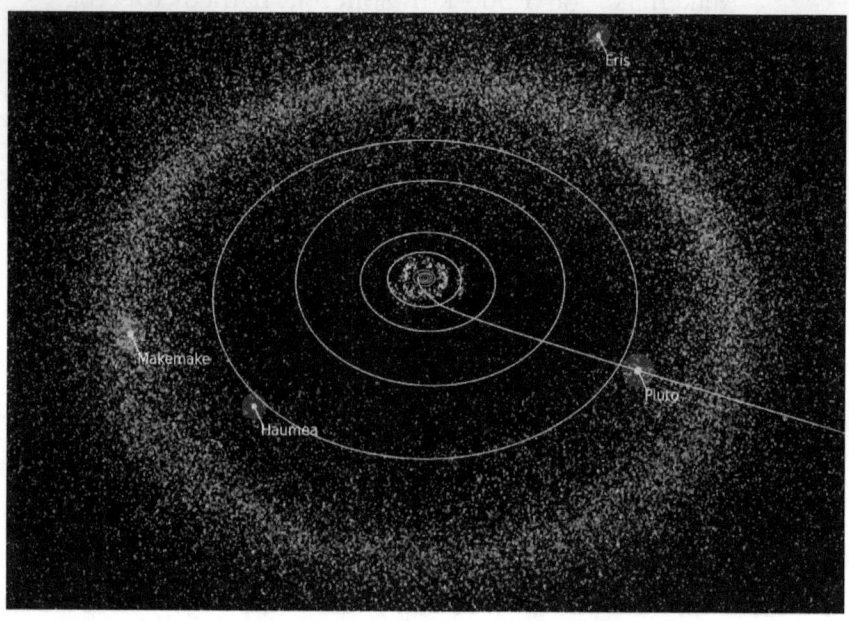

Distribution of known Kuiper Belt objects. Labeled ones are Eris (upper right), Makemake (left) Haumea (lower left), and Pluto (center right). The line represents the flight path of NASA's New Horizons spacecraft, which flew by Pluto on July 14, 2015. Image by NASA/JHUAPL.

Washington. "This represents just the tip of the iceberg," he said. "I think we'll find several hundred to a thousand." Since then, the University of Hawaii has been at work building a new, improved version called PS2, with the two telescopes expected to work in tandem, which should provide even better resolution.

Another telescope, the Large Synoptic Survey Telescope, is being planned for a Chilean observatory. It should be able to push the observable size limit down to 50 to 100 kilometers once it becomes operational in 2020. "We would expect about 30,000 objects," says one of the project's leaders, Lynne Jones of the University of Washington.

But astronomers aren't content with counting planetoids. They also want to know their orbits. And based on these, what they've found are two basic groups of Kuiper Belt objects. One, sometimes called the cold, classical Kuiper Belt (or simply cold Kuiper Belt), lies in the same plane as the rest of the Solar System, mostly about 4 to 5 billion miles from the Sun. These objects, which comprise about half of those discovered to date, aren't "cold" because they're more frigid than the rest, but because their sedate, in-the-plane orbits are dynamically calm. Other Kuiper Belt objects zip high out of the plane, like satellites orbiting above the Earth's poles. These are dynamically "hot," and are referred to as the hot Kuiper Belt. They're particularly interesting because, as we shall see, they probably didn't form in their present orbits. "There's currently a lot of debate about what happened in the early history of the Solar System," Jones says. "We used to think things were relatively quiet, but we can see that this is not the case. They are dynamically much more excited than we would expect."

A Belt That Shouldn't Exist

Figuring out what it is that kicked so many "hot" Kuiper Belt objects out of the plane of the Solar System isn't the only problem facing theorists. Kuiper's original prediction notwithstanding, modern scientists have struggled to explain why there's anything out there at all. "Classical planet-formation models have a hard time 'growing' big objects at this distance from the Sun," said Alex Parker, then a graduate student in planetary science at the University of Victoria, at an October 2010 meeting of the American Astronomical Society's Division of Planetary Science (DPS). "Unless you start with a lot more mass in the primordial disk and then get rid of it later, these models can take longer than the age of the Solar System to grow the kinds of objects we see."

One problem planet-formation models have is that they start by postulating that the Solar System began as a disk of dust and gas circling the sun like a sombrero. But that protoplanetary disk didn't extend outward forever. Most likely, it stopped somewhere around 30 to 35 astronomical units (AU) from the Sun, not far beyond the orbit of Neptune (which lies at 30 AU).[1] Unfortunately, the Kuiper Belt doesn't seem to know the theory. Eris swings in an elliptical orbit ranging from 38 to 98 AU. Makemake is between 45 and 55 AU. Haumea ranges from 35 to 52, and Sedna's minimum distance is 76 AU, while its maximum is more than 900 AU.

Perhaps the protoplanetary disk was simply larger than thought, but that leads to the problem alluded to by Parker, which is that even with thousands of objects in the Kuiper Belt, there just isn't a lot of mass: no more than one-tenth that of Earth by one estimate. Even if continuing discoveries raise that figure, the total mass is still far too small for a protoplanetary disk dense enough for these objects to have formed that far out. Under traditional models, the Kuiper Belt either shouldn't be there or should be a lot more densely packed.

A few years ago, these problems led theorists to propose that

---

[1] An astronomical unit is the distance from the Earth to the Sun, approximately 93 million miles.

today's Kuiper Belt objects must have formed much closer to the Sun, possibly in the same zone as Jupiter, Saturn, Uranus, and Neptune. Once formed, they were flung outward by dramatic shifts in the orbits of these planets, most importantly Neptune. That theory, known as the Nice model (because it was developed, in part, by scientists from Nice, France), quickly became a leading explanation for the Kuiper Belt. "In some of these simulations, they even suggest that Uranus and Neptune switched places," Parker said.

There's just one problem. Parker has found at least one class of Kuiper Belt objects for which this simply couldn't have happened. These objects are binary pairs, like Pluto and its moon Charon. Such pairs are common in the Kuiper Belt—estimates are that at least 30 percent of objects in the cold, classical Kuiper Belt exist in pairs or larger groupings. But Parker found seven such pairs that are so widely separated they circle each other in a very slow waltz, with orbital periods from four to 17 years. "They're delicate," he said, "weakly bound." Any close gravitational encounter with a giant planet would have pulled them apart and sent them in different directions. "They would not be there today if [they] were ever hassled by Neptune," Parker said.

Nice-model theorist Alessandro Morbidelli of the Laboratoire Cassiopee, Nice, France, agrees. "The Parker result calls for at least some tweak in the Nice model," he says. "Either the cold Kuiper Belt (or part of it) formed *in situ*, or it was pushed to its current location by a mechanism less violent than envisioned."

Stephen Tegler, a planetary scientist at Northern Arizona University, concurs. "We have to come up with a way to tweak the Nice Model, or they formed in situ," he says.

Collisional Cascade

Enter Mike Brown, a cheery astronomer from California Institute of Technology who looks a bit like a young Bill Gates. Brown's Twitter handle is "plutokiller" because of his role in kicking Pluto off the list of planets. But he's also co-discoverer of Eris, Sedna, Makemake, Haumea, and nearly a dozen others, and author of *How I Killed Pluto and Why It Had It Coming* (Spiegel & Grau, 2012). And like just about everyone studying the Kuiper Belt, he too has come up with surprises to upset the theoreticians' apple carts.

In his case, the problem stems from the newly found worldlets' densities. Traditional theories say that planets condensed out of the primordial disk bit-by-bit in an incremental process that saw microscopic particles coalesce into ever-larger ones until eventually they formed planets. But as we noted before, the solar disk should have been getting rather tenuous out toward the Kuiper Belt. "To get something that large, you would have to have had to accrete from a very large swath of the outer Solar System," Brown says. This should have averaged out any random variations in the nebula's composition. "You would think they would be some of the most uniformly comprised objects in the Solar System."

Pluto fits this model perfectly. Its density is about 2.0 grams per cubic centimeter, indicating that it's composed of about a fifty-fifty mix of rock and ice, just about what theoreticians would expect. But not everything is like Pluto.

The first discrepancy came from Haumea. It's not merely elongated, it's flattened. Earlier I called it football-shaped, but that was a simplification. Brown calls it a football "that's a bit deflated and stepped on." It's also spinning end-over-end so fast it completes a full revolution every four hours. Small asteroids often look and spin like that, but Haumea is big enough it wouldn't fit between San Diego and Vancouver, Canada. And it turns out to have a density of about 3.0 grams per cubic centimeter. "That's 100 percent rock," Brown says.

As far back as 2006, Brown suggested that Haumea might be the core of a larger planetoid that got clobbered early in its early history. A glancing blow would not only have set it spinning but knocked off its icy outer mantle. Supporting this is the observation that it has two tiny moons that are almost pure ice. Could these be chips off the original ice block? A year later,[2] Brown's team reported additional evidence: the discovery of five other fragments whose orbits indicate they were blown away from Haumea near the dawn of the Solar System. The fragments (the largest of which might be 400 kilometers across) appear to be made of nearly pure ice—just right to be more of Haumea's missing mantle.

One anomalously dense world created by a mantle-busting collision could be a fluke. But as more and more numbers come in, the densities of other large Kuiper Belt objects appear to be all over the place, ranging from well below 1 gram per cubic centimeter (the density of frothy ice) to nearly as dense as Haumea. Brown suggests this means these objects didn't form by gradual accretion. Instead, he thinks they originated from large chunks, several hundred kilometers in diameter—large enough to have already differentiated into rocky cores and icy

---

[2] *Nature*, 15 March 2007, pp.294-296.

mantles. These primordial objects, sometimes called planetesimals, then bashed together in big collisions, a process he calls "pyramidal growth."

That way, even the largest could have formed in only a few steps. Each collision would have blasted debris into space—the source of today's smaller Kuiper Belt objects, like those associated with Haumea. Sometimes, two ice-depleted objects would have collided, producing a larger ice-depleted object. Other times, icy chunks of a one-time mantle would merge into larger, rock-depleted worlds. Still other collisions would produce Pluto-like mixes of rock and ice.

This only works, however, if the main growth came from collisions between large objects rather than stepwise accretion from small ones. Otherwise the law of averages would make everything more-or-less identical.

Further evidence against traditional incremental-growth collisions comes from Parker's binaries. They are so far apart that it's not merely a close encounter with Neptune that would have disrupted them. They'd also have been knocked out of each other's orbits if they'd been pelted by too many objects only a kilometer or so in diameter: a strong indication that there were never many such objects around—or, in other words, that Kuiper Belt objects started out big.[3]

Rapid Condensation

[3] The same may also have applied to the asteroids. When planetary scientists plot their diameters, they find a hard-to-explain excess at diameters of about 100 kilometers. In a 2009 paper in *Icarus*, a team led by Morbidelli argued that reproducing this via asteroid formation models requires starting with planetesimals of at least 100 kilometers in diameter. See Morbidelli, et al, "Asteroids were born big," *Icarus* 204 (2009) 558–573.

All of this is interesting but begs the question of where the first round of objects came from. Somehow, the ancestral Kuiper Belt had to have jumped all the way to planetesimals without going though intermediate steps.

The existence of Kuiper Belt objects, however, isn't the only hint that something might be amiss with traditional stepwise-growth models. It's not that the conventional theory isn't appealing. It's nice, orderly—tidy, in fact. Microscopic bits of dust and ice coalesce, first into dust bunnies, then into tennis balls, beach balls, etc., until you wind up with objects large enough to be called planetesimals. But there's always been a problem. It's easy to get from dust cloud to dust bunnies. It's easy to model what happens to planetesimals once they reach kilometer size. But there's a no-man's-land between, Brown says, in which the models just won't let the midsize objects accrete. Instead, they quickly lose orbital speed from gravitational interactions with another component of the solar disk: gas. You could think of this as a type of gravitational drag whose effect is that small objects spiral into the Sun more rapidly than they can merge into larger bodies.

It's one of those things that tends to inspire scientific hand waving. *So there was this disk of gas and dust,* people want to say. *It formed a bunch of little clumps, and somehow* (enter the hand-waving) *these stayed around long enough to merge into things the size of football stadiums. These then merged into planets, moons, the Earth, etc.* That step in the middle has been the elephant-in-the-living-room of planet-formation models. "It's a mystery nobody wanted to think too hard about," Brown says.

The answer, he and Parker believe, lies in an emerging set of planet-formation models. Like their predecessors, they start with a dusty solar disk, but instead of focusing solely on stepwise accretion, they include the effect of eddies, vortexes, and turbulent flows in the early disk. Brown compares these turbulences to brooms sweeping dust particles together. By creating dense zones in the solar disk, they make for a shortcut process of planet formation in which "big hunks" fall together without need for intermediates. "The first things [to be formed] are 100, 200, and 300 kilometers, instead of millimeters," he says.

What we then have is a Kuiper Belt in which everything originated as rapidly condensing planetesimals. Today's big objects are the result of Brown's collisional cascade. Small ones are fragments blown away by collisions. Mid-sized ones are survivors from the original population of planetesimals.

But we still have problems. To begin with, the only collision we have evidence of is the one that stripped Haumea of its mantle. The ability to trace fragments from it clearly shows that it occurred out in the Kuiper Belt, but it's actually rather startling that there's any evidence of any such collision, ever. The Kuiper Belt is a gigantic zone, covering far more volume than all of the inner portions of the Solar System, combined. It's hard to imagine how the objects we see out there ever managed to hit each other often enough for Brown's collisional cascade to have occurred.

And we still have a missing-mass problem. The turbulence models still require a minimum density for Brown's "big hunks" to condense. As before, there either should be more (or bigger) objects out there…or none at all.

Possibly, some of the missing objects were pulverized by giant collisions, reducing them to dust that fell into the Sun by gas drag. But Morbidelli doesn't buy it. "You cannot hope that big objects disappear into dust by collisions," he says. "They are too big to destroy completely." Instead, he believes that the Nice model is still the best explanation. In fact, he thinks Brown's collisional cascade adds considerable support for it by indicating that Kuiper Belt objects originated in a more closely packed region, where there was more opportunity for mantle-busting collisions. Furthermore, the Nice model solves the missing-mass problem because most encounters with Neptune would not have been gentle. In the process of being flung outward, computer models show that only one object in a thousand would actually wind up in the Kuiper Belt. The rest are ejected into interplanetary space. "That is exactly what we need," Morbidelli says.

That way of viewing it gets us all of the big, collision-produced objects and delivers them nicely to their present locations. But Parker's binaries are another story. "His result implies that at least part of the cold Kuiper Belt never experienced encounters with Neptune," Morbidelli admits.

Perhaps, he says, the hot and cold Kuiper belts were produced by different mechanisms. This would mean the hot Kuiper Belt objects were produced by encounters with Neptune, but not gravitationally violent enough to have torn apart Parker's binaries.

Morbidelli and Harold Levison of the Southwest Research Institute in Boulder, Colorado, suggested one way this might have occurred as far back as 2003: an orbital resonance with Neptune.[4] Such resonances are common in outer-planet moon

---

[4] Levison & Morbidelli, *Nature*, 27 November 2003, pp. 419-421.

systems, where, among other things, they create intriguing features in Saturn's rings (see Chapter 3). In this case, the binaries would have formed farther out than Neptune's original orbit, then been slowly nudged even further out by a recurring gravitational force from Neptune, as Neptune itself migrated to its present position. And that's not the only possibility, because we don't know exactly how Neptune reached its modern orbit. "Many variants are possible," Morbidelli says.

But again we have a missing-mass problem because now we can't use Neptune to toss most of the initial population of objects off toward the stars. "This is a great conundrum," Morbidelli admits. To date, there is no clear solution. One possibility is that the solar disk really did extend into the Kuiper Belt with enough density for planetesimals to form, at least as far out as Parker's binaries. Perhaps that far out, it contained a lower fraction of dust, altering the process of planetesimal formation so that the rapid-condensation turbulence process didn't occur. That might mean that while, closer in, most of the dust became big planetesimals, further out most of it simply spiraled inward by gas drag. If so, Morbidelli says, it's possible that a few large planetesimals—today's cold, classical Kuiper Belt—could have formed in the 35 to 45 AU range, with the missing mass having disappeared simply by gas drag that eventually dumped it into the Sun. In addition to solving the missing-mass problem, this theory also has the advantage of suggesting that the existence of Parker's binaries means exactly what it appears to mean: that they formed in situ.

But, Morbidelli notes, at present all of this is mere speculation. "To go beyond will take a lot of time."

RICHARD A. LOVETT
Rogue Planets

From a science-fictional perspective, this provides wonderful fodder for stories. In his classic novels *Gateway* and *Beyond the Blue Event Horizon*, Frederick Pohl postulated a CHON (carbon, hydrogen, oxygen, and nitrogen) food factory on a proto-comet far out on the edges of the Solar System. It lay in the Oort cloud, a shadowy zone that might be the birthplace of comets, nearly a light-year from the Sun. But the Oort cloud is a long way out. The same resources should be findable in the Kuiper Belt (which wasn't discovered until 1992, more than a decade after Pohl wrote his stories).

Furthermore, many science fiction stories have envisioned asteroid-belt civilizations. The Kuiper Belt might provide something similar, not too impossibly remote, especially for super-loner types who think the asteroid belt is too close to civilization.

What we know about the Kuiper Belt's formation also raises other interesting science-fictional possibilities. In a 2011 paper in *Nature*, a team led by Takahiro Sumi of Osaka University, Japan, announced the discovery of free-floating Jupiter-sized worlds in interplanetary space. Using a technique called gravitational lensing, in which otherwise too-faint objects are revealed by their gravitational effect on light passing by them, the scientists spotted ten such planets in a single search—enough to conclude that rogue Jupiters, unbound to any star, might outnumber stars by a factor of two to one.[5] These giant planets would have gotten tossed away from the stars of their birth by processes similar to those that, under the Nice model, may have hurled

---

[5] Sumi, et al, *Nature*, 19 May 2011, pp. 349-352.

myriads of smaller bodies into interstellar space. The bottom line is that planets and stars need not go together. And while rogue Jupiters probably wouldn't be particularly habitable, rogue Plutos or Sednas might be—especially because there are arguments that some of these bodies might even have liquid water deep in their mantles.

Science fiction has long been fascinated with generation-ship stories in which colonists head for the stars in giant spaceships that will take thousands of years to complete their voyages: ships that become entire worlds to their crews. But maybe we don't need such spaceships. Maybe we don't even need to go to the stars. Maybe everything we need is out in interstellar space, waiting for us to hitch a ride. Or maybe these worlds are already teeming with aliens who've been so long aboard that they see no need for stars or the dangers that go with them. And that's enough maybes for a whole suite of stories.

Meanwhile scientists are struggling to learn more about the Kuiper Belt.

As this book is going to press, NASA's New Horizons spacecraft has flown by Pluto and given us our first close-up look at two Kuiper Belt objects: the no-longer-planet Pluto and its moon Charon. If you were reading the news in 2015, you'll know that the mission returned some dramatic photos, proving that Pluto was nothing like we'd imagined, with giant mountains, active geology, and blue skies. But one of the other things that mission will eventually allow astronomers to do is to tally the numbers and sizes of craters on both worlds, providing important information about the types of objects that have battered them. If there are lots of small craters, then the Kuiper Belt is full of kilometer-sized objects, even if we don't know why. If there are mostly giant impacts and a relative handful of small ones, that's further confirmation that the Kuiper Belt's progenitor planetesimals were indeed large. Ultimately, Brown says, once planetary scientists have combined crater counting studies with the latest modeling results and the growing tally of known Kuiper Belt objects, "I think we'll know the basic outlines [of Kuiper Belt history]."

And that, he says, could have broad implications, including shedding more light on our own planet's origins. "We could well be completely rewriting how planets form," he says.

Author's Note

The Kuiper Belt is one of the most enticing arenas for Solar System research because the more we learn about it, the more intriguing it appears. One of the most exciting opportunities of my science-writing life thus came in July 2014, when I was invited to join the press corps at mission headquarters as New Horizons made its historic flyby of Pluto.

I was there because I'd convinced an editor that New Horizons would be the last great voyage of discovery of my lifetime—not just because we'd be getting our first close-up look at a Kuiper Belt object (Pluto), but because it was the first time since 1989 (during the Voyager flyby of Neptune) that we'd gotten a first close-up look at any large chunk of Solar System real estate. Maybe someday we'll do a flyby of Eris or some other large Kuiper Belt object, but it won't be soon.

What we discovered was nothing short of amazing. Pluto proved to be a world with blue skies (who would've thought?), ice volcanoes thousands of feet high, and a startlingly active geology. In a single 24-hour flyby, New Horizons took so much data it required 16 months to beam it all back to Earth…and who knows how many years afterward for it all to be fully analyzed. One of the reasons there is no story about Pluto in this book is that it would be guaranteed to be out of date by the time you read it.

Instead, lets come back to Earth for a topic I'd never heard of before I encountered it in a scientific symposium. It's a fascinating example of what scientists can do when they have exquisitely sensitive instruments, a lot of data, and enough computing power to tease out details nobody would ever have thought possible.

Forensic Seismology

The Big Science of Minor Shake-Ups

On August 19, 2000, a dozen people were on a family reunion, camping beside a river in southern New Mexico. What they saw and heard isn't entirely known, but we can imagine much of it. Most likely, there was a noise like a clap of thunder, originating from a source 200 yards from the campsite. Rocks and debris rained from the sky, but nobody was seriously hurt. Then perhaps the dust cleared enough to reveal a crater twice the volume of an Olympic swimming pool where, moments before, there had been nothing but open ground.

Perhaps the campers gaped in wonder, trying to figure out what had happened. Perhaps they hit the dirt, like soldiers under fire. Maybe the concussion left them dazed, or maybe they heard the ominous hiss of *something* escaping from the pit. Possibly, they thought a volcano had chosen that moment to announce its presence and were waiting for the first sight of lava. Most likely, few of them realized that a weakness in a buried natural gas pipeline had caused it to blow out, and that the real danger was wafting toward them on the breeze.

Seconds ticked by and nothing else happened...nothing, until their world erupted in flame, accompanied by an apocalyptic roar that went on and on. In the crater, the temperature hit 2,200 °F. At the campsite, it wasn't that hot, but it was hot enough. Some of the campers lived long enough to be rescued an interminable hour later, but all subsequently died of their burns.

Many people were involved in the ensuing investigation, but one of the more unusual was Keith Koper, an assistant professor of geophysics at Saint Louis University. Koper was a seismologist, called in to help reconstruct the events that occurred during the blast and the resulting inferno. While other people dug through debris, Koper poured over vibrations recorded on seismometers up to 110 miles away.

Most people associate seismology with earthquakes—as well they should, because seismometers are the key instrument for studying temblors. Seismometers, however, are simply very sensitive needles designed to record vibrations in the ground. They aren't picky about whether the source of the vibration is an earthquake, a volcanic eruption, or the hoof beats of a wild horse. A team of researchers studying earth movements in the Andes Mountains, for example, was baffled by vibrations that were only observed at night. Earthquakes can occur at any time of day, so it appeared that their instruments were picking up something else. Eventually, the scientists realized they were watching the trucks of smugglers, ferrying clandestine cargoes across a remote section of the Chilean/Argentine border. After studying the seismometer traces for a while, the scientists could even figure out the direction in which each truck was going, said Terry Wallace, now principle associate director for global security at Los Alamos National Laboratory.[6]

Decades ago, seismologists would have been frustrated by such human-caused interferences. Then, during the Cold War, defense experts realized that seismometers could be harnessed to monitor the other side's nuclear tests. Even if the tests were underground, the delicate needles of seismometers could detect them from thousands of miles away. It was the birth of forensic seismology.

Seismology is a science that can be as simple or as complex as you want to make it. In the vicinity of a major earthquake, it's pretty simple. There, everything—and everybody—is a seismometer, jounced around at the whim of powerful forces. In 1994, for instance, a friend of mine was living near the epicenter of Southern California's Northridge quake, which measured at magnitude 6.7. She first became aware of the earthquake when she awoke in midair between her bed and the floor, having been slammed by a shockwave that scientists later determined to have produced an acceleration more than 1.6 times the force of gravity. In the technical argot of seismology, such up and down vibrations are called Rayleigh waves, which ripple along the Earth's surface like waves on a pond. But my friend didn't need an expert to tell her what had happened. To her, it felt like she'd been hit from below by a giant fist.

Seismometers can be designed to record ground motion in any of three directions: up-and-down (as in the jolt that threw

---

[6] I first heard this and many other anecdotes in this article at a 2002 meeting of the American Geophysical Union, but while they are old, they remain classics of the field.

my friend out of bed), side-to-side, or forward-and-back. These motions can be quick jerks or prolonged shaking lasting for several minutes. They can be bursts of high frequency seismic noise or slower, steadier oscillations.

Seismology gets complex when you start using the direction of motion and the frequency and duration of vibration to deduce the cause of the tremor. Explosions have seismographic signatures that "look" quite different from earthquakes, and the sluggish, stomach-rumbling vibrations from the movement of lava beneath a volcano don't look anything like what you'd get from a meteor impact.

Seismology becomes even more complex when banks of instruments are used in array to record the same event from different distances and directions. Currently, there are well over 10,000 permanently installed seismometers scattered around the globe, many of which upload their readings to the Internet, practically in real time. A big earthquake will show up on all of them; a technician's incautious door-slamming will appear only on the one being serviced. Events of true forensic interest fall somewhere between the two extremes, but to triangulate on the location of a seismic event, it needs to be strong enough to be observed by at least three seismometers. The more instruments that record it and the more broadly scattered they are, the more accurately the source can be pinpointed.

The Seismic War on Terror

Seismic waves travel at speeds of 2,000 to 16,000 miles per hour, depending on the type of rock through which they are being transmitted. That's fast enough that big events, such as major earthquakes, are felt by seismometers on the opposite side of the globe within hours, sometimes before the first scrambled communications arrive from the stricken area (though Twitter is changing this). For human-caused events, such as military actions, industrial accidents, or terrorist attacks, however, there are generally faster ways to know what happened. In these cases, forensic seismology is more often useful after the fact, when investigators turn to seismographs to see what these records might reveal.

When Pan Am Flight 103 was brought down by a terrorist's bomb in 1988, over Lockerbie, Scotland, for example, nearby seismometers recorded the stricken airliner's impact with the ground. In fact, seismologists identified six major impacts, indicating that the airplane had broken into that many large pieces before it hit. This was already evident from the pattern of debris at the crash site, but the seismic data allowed the scientists to do something more: they could calculate the kinetic energy of each of the six pieces. Comparing this to the weight of the fragments recovered at the site allowed a simple calculation of each piece's speed at the time of impact—information that came into play in reconstructing the explosion that ripped the jetliner apart.[7]

Similar seismic detective work was done regarding the sinking of the Russian submarine *Kursk* during naval exercises on August 12, 2000.[8] When Wallace and Koper heard the news, they rushed to download data from seismometers in Norway,

---

[7] David A. McCormack, *Eos. Trans. AGU*, 83(47), Fall Meet. Suppl., Abstract S11D-02, 2002.

[8] Wallace & Koper, *Eos. Trans. supra*, Abstract S11D-05, 2002; also, Sid Perkins, "Explosions, Not Collision, Sank *Kursk*," *Science News* (online), Jan. 27, 2001.

Sweden, Finland, and Spitsbergen (an arctic island north of Norway). Underwater explosions don't directly shake up the Earth's crust in the same way as earthquakes, bomb tests, or airline crashes, but they do produce shock waves that slam into the seabed, transmitting energy to the rocks below. When the *Kursk* sank, the Scandinavian seismometers recorded the event as equivalent to a magnitude 4.1 earthquake. These seismometer readings not only allowed U.S. observers to pinpoint where the submarine went down but showed that the main explosion had been preceded, two minutes and 15.75 seconds earlier, by a smaller seismic event of about magnitude 2.2.

Based on the physics of underwater explosions and other researchers' tests of the efficiency with which this energy is transmitted into the seabed, Wallace and Koper were able to determine how deep the submarine was below the surface at the time of the accident. That's because underwater explosions produce bubbles of hot gas, which oscillate as they rise to the surface. The frequency of these oscillations depends on the size of the explosion and the depth at which it occurred, allowing the scientists to determine (despite Russian claims to the contrary) that the cause of the accident was an explosion, and that it had occurred at a depth of 83 meters. The first blast carried a power equivalent to approximately 250 kilograms of TNT, while the second was equivalent to somewhere between 4,000 and 7,000 kilograms. Because the first explosion was about the size of the charge on a torpedo—and because the *Kursk* was known to be conducting a torpedo exercise at the time it sank—Wallace and Koper posited that it was a torpedo misfire. The second, larger one, they suggested, occurred when fire from that accident detonated additional warheads.

Shortly after the sinking, the seismologists noticed many small events with magnitudes between 1.25 and 1.86 on the Richter scale. These appear to have been depth charges dropped by the Russians to discourage other nations from sending scuba divers to steal the secrets of their ill-fated submarine. The seismic readings were so precise that the researchers could even track the speed and course of the naval vessel used to lay down the underwater barrage. Based on the observed seismicity, it appears that the Russian depth charges were detonated at random depths, ranging from ten to 80 meters (the submarine was trapped at 107 meters). The scientists calculated that they carried charges equivalent to 112 kilograms of TNT—not too far off from what military experts believed to be the Russian munitions' true explosive charge.

Ever-vigilant seismometers were also on duty during the 9-11 terrorist attack on the World Trade Center. Because tall buildings are built on deep footings, energy from the airplane impacts shivered down the spines of these structures directly into the bedrock, allowing them to be detected by seismometers several hundred kilometers away.[9] The collapses of the buildings also produced seismic traces so detailed it was possible to identify the crash of each floor as the buildings pancaked downward in a series of rat-a-tat-tat impacts, each stronger than the one before.

The Forensics of Global Climate Change

Not all uses of forensic seismology are as morbid as tracking the collapse of the World Trade Center. More than a decade ago,

---

[9] Won-Young Kim, *Eos. Trans., supra,* Abstract S11D-03, 2002.

for example, Vera Schulte-Pelkum of the University of Colorado examined Depression-era seismic records in search of the answer to an important question about global climate change: have the great storms of the North Atlantic Ocean become rougher and more frequent in the past century?[10]

Today, we monitor such things via wave-height sensors in ocean buoys, but those instruments haven't been used for all that many years. Seismograph readings date back to the 1930s. Using these, researchers such as Schulte-Pelkum have been able to tease out the faint, rhythmic signals of surf crashing on far-distant shores. That's not quite the same thing as direct measures of wind speed, rainfall, and wave height, but it's still a nice proxy for storm intensity. And even Depression-era seismometers were such remarkably delicate instruments that this data appears to be adequately preserved in their traces.

Waves aren't the only oceanic events that can be observed by seismology. In late 2000, a seismic monitoring network in Polynesia detected several episodes of strong activity coming from the Southern Ocean, producing signals that resonated at about four cycles per second.[11] At first, geophysicists thought the signals came from volcanoes—not an odd presumption, because underwater eruptions have previously been discovered in this manner.

Soon, the source was narrowed to the Ross Sea, which lies near America's McMurdo Station in Antarctica. One of the

---

[10] Schulte-Pelkum, et al, *Geochemistry, Geophysics, Geosystems*, March 2004.

[11] Jacques Talandier, et al, *Earth and Planetary Science Letters* (2002) pp. 519-534.

world's largest active volcanoes, Erebus, is nearby, so the researchers figured that they were simply seeing that mountain's latest activity. But something was wrong: the volcano seemed to be moving. Puzzled, the scientists tracked the signal for about 125 miles, at which point it disappeared. What, they wondered, could be moving across the Ross Sea at about 3.5 miles per day, producing volcano-scale seismic rumblings along the way?

It didn't take long to come up with an answer. Earlier that summer, a big chunk of the Ross Ice Shelf had broken loose in a swarm of gigantic icebergs. The iceberg that their instruments appeared to be tracking was 80 miles long and the noise appeared to come from water-filled cracks that vibrated, either as the berg collided with neighboring behemoths or as its base dragged along the Ross Sea's shallow floor. Whichever of these was the cause of the noise, the moment the iceberg moved into deep water, the signal disappeared, confirming the scientists' belief that the large chunk of ice was indeed what their instruments were picking up.

Rumblings in the Air

Forensic seismology has two sister sciences, hydroacoustics and infrasound. Hydroacoustics involves listening to low-level noise in water, typically in the ocean. It overlaps seismology when these noises are transmitted into the sea floor, as in the sinking of the *Kursk* or the sounds of Antarctic icebergs, but it's also possible to conduct underwater investigations based solely on hydroacoustics. (It's also a military technology that played a major role in the Cold War, glamorized in Tom Clancy's submarine novel, *The Hunt for Red October*.)

HERE BE THERE DRAGONS

Infrasound occurs in air. It's simply sound, pitched at frequencies too low for human hearing. Technically, it's any sound whose frequency is below 20 Hz (cycles per second), but forensic applications generally examine frequencies in the range of 0.1 Hz to 4 Hz. This is the type of sound which, if it's strong enough, you feel as a palpable thump. But delicate instruments can detect it at far lower intensities. Called microbarometers, they measure changes in air pressure as infrasound waves pass across them, and are so sensitive they could detect the altitude-related air pressure difference between the top and bottom of a sheet of paper.

One use of infrasound is to detect aboveground nuclear tests—not of megaton-scale weapons (easily detectable by many other means), but of the smaller tactical weapons most likely to proliferate. Thus, the Comprehensive Nuclear Test Ban Treaty Organization is constructing a 60-station global infrasound monitoring system, at an estimated cost of $35 million. Several stations are already operational, and—not surprisingly—they've detected infrasound from a sizeable number of non-nuclear events, opening the door for uses of infrasound that have nothing to do with nuclear monitoring.

Douglas Revelle, a scientist from Los Alamos National Laboratory who died in 2010, was an expert in the collisions of the Earth with large meteors, often referred to as bolides. The most famous of these is the 6-mile diameter rock that appears to have struck Mexico 65 million years ago, widely believed to be the event that killed off the dinosaurs. But many smaller rocks hit the Earth each year, sometimes striking ground, but more often exploding in the atmosphere with resulting claps of infrasound.

RICHARD A. LOVETT

Observing bolides by infrasound isn't new. In 1908, an asteroid estimated to be about 125 feet in diameter exploded above the Tunguska region of Siberia, leveling forest for 25 miles in all directions and producing an infrasound signal so strong that even the crude barometers of that era tracked its pulse through three full circuits of the globe.

The Tunguska bolide hit with an energy equivalent to a 10-megaton bomb—truly devastating if something similar occurred in a populated area. One use of the infrasound network is to collect statistics on the size of smaller bolides in an effort to refine our estimates of how often Tunguska-scale events might be expected to occur. In studies carried out over the course of several decades, Revelle found that the upper atmosphere is repeatedly peppered by large explosions, ranging from less than a thousandth of a kiloton (the detection limit of the infrasound sensors) to the largest detected in his lifetime, which struck in 1994 and packed a whopping 200 kilotons.[12] (More recently, the Chelyabinsk asteroid, which blew out thousands of windows in Russia but miraculously didn't kill anyone, came in at 500 kilotons.)

Not surprisingly, the smaller the energy level, the larger the number of events. In a typical year, there are about 30 events in the 0.1-kiloton range and seven in the kiloton range. Tunguska-scale events have been estimated to occur about once every 1,000 years, though the Chelyabinsk airburst has caused scientists to wonder if they might be more frequent than

---

[12] To put this in perspective, an 8-foot rock would produce about a 1-kiloton blast, depending on speed and trajectory.

43

previously thought.[13]

Infrasound can also be used to listen to volcanoes, each of which has its own infrasound voice, says Milton Garces, an infrasound researcher at the University of Hawaii. "Some are operatic, and some are more like a colicky child." Garces' goal is to learn these mountains' infrasound languages, so he can distinguish the sounds of contentment from those of threat. "All volcanoes have a gas problem," he adds with a grin. "Happy volcanoes release gas steadily, purring or rumbling. Unhappy volcanoes release gas explosively." This is important, says Douglas Christie, of Australian National University, because volcanic ash is abrasive to the engines of jetliners, and infrasound might serve as a means of locating remote eruptions and issuing prompt warnings.

Another use of infrasound is for avalanche warnings. Anyone who's ever heard one knows that an avalanche produces not only an audible roar but a chest-shaking vibration of palpable infrasound. Alfred Bedard, Jr., of the National Oceanic and Atmospheric Administration's Environmental Technology Laboratory in Boulder, Colorado, has noted that simple detectors could be placed near known avalanche paths and linked to electronic traffic signs. When an avalanche occurs, the sensors could activate the signs to stop motorists at a safe distance. More sensitive detectors, located at greater distances, could count the number of avalanches in an entire mountain range and help avalanche forecasters improve warnings issued for backcountry skiers.

---

[13] P.G. Brown, et al, *Nature* 503, 14 November 2013, pp. 238–241.

RICHARD A. LOVETT

Bedard has also worked with infrasound as a means of fine-tuning tornado warnings, beginning with a pilot project designed to monitor parts of Colorado, Kansas, Nebraska, Oklahoma, and Wyoming with a three-station network. (As with seismology, three infrasound stations are needed to triangulate on the source of a signal.)

Tornadoes emit infrasound when the funnel cloud is stretched. This causes it to vibrate radially, at infrasound frequencies. The larger the tornado, the slower is the frequency of vibration. A funnel cloud that's a quarter-mile in diameter (a big tornado) will vibrate at about 1 Hz. Small funnels, too tiny to show up on Doppler radar, vibrate at 5 Hz to 10 Hz. The goal is to track and count tornadoes with greater precision than current methods permit.

It may also be possible to use infrasound for earthquake warnings. Bedard has reported that a study in Armenia has examined infrasonic waves that appear to be emitted as precursors to earthquakes. These waves propagate outward into the atmosphere, where they affect the ionosphere in ways that can be detected by radar.

Because infrasound involves low frequencies, it takes large-scale events to produce it. This limits infrasound's forensic uses, but doesn't mean that they're nonexistent. Space launches and re-entries can be detected from more than a thousand miles away, and following the explosion of the Space Shuttle Columbia in early 2003, infrasound researchers quickly began poring over their recordings in search of clues to what might have gone wrong.

Other infrasound sources range from the mundane (traffic noise on busy freeways) to the exotic (the Northern Lights).

Important human sources include the oil-well fires set by the Iraqi military as it retreated from Kuwait in 1991, sonic booms, and chemical explosions (e.g., a fire at a fireworks factory). Overall, says Christie, there are at least 40 known sources of infrasound, "and we're going to have a lot more."

Pipeline Revisited

So what happened during that pipeline explosion in New Mexico? As luck would have it, two arrays of seismometers had been deployed in the area for short-term geological studies, and more than a dozen of these instruments recorded usable data about the explosion.

Koper's first observation was that the seismic signal lasted for more than an hour. During that time, the fire showed up as an easily noticeable vibration that quit at exactly the time when the pipeline company said the blaze had been extinguished. Nobody had really doubted the company's records, but the attorneys on Koper's side were happy to have independent verification.

The real interest was in the seismic details. These showed several distinct stages in the first minute of the event. Stage one was the pipeline blowout. It appeared in the seismographs as an explosion-style jolt. Stage two was a continuous high-frequency vibration, much weaker than the roaring from the ensuing fire. This, Koper concluded, must have come from gas leaking from the pipeline, scouring the earth as it jetted out of the hole. The fire had not yet started. Then, 24.0 seconds later, there was a second bang, followed by the roar of flames. This, Koper concluded, had to be the explosion that ignited the fire. It was followed, 17.8 seconds later, by a third explosion, probably from

46

the ignition of a pool of gas that had not been ignited by the initial spark. At this point, the roaring stepped up a notch, indicating that the pipeline had been fully converted into a gigantic torch. But the initial scouring signal remained discernable: a high-frequency hiss, masked but not obscured by the louder signal from the conflagration itself.

When the pipeline company finally began to get the fire under control, it didn't shut off all at once. Rather, it diminished in a series of steps, as valves were shut down one by one, reducing the amount of gas feeding the flame. The last of these marked the time when rescue workers were finally able to approach the victims.

All of this proved highly relevant to the ensuing legal settlement. Because some of the victims survived long enough to be transported to the hospital, it is arguable that they might have survived had rescue workers been able to get to them sooner. Some pipelines have automatic shut-off valves that quickly clamp down in the event of a leak. This one, Koper says, did not. The seismometer readings provide independent confirmation of company records indicating that in the absence of valves, it took an hour to shut off the gas.

More important, though, is the 24-second gap between the blowout and the primary ignition. This, Koper says, implied two important things. The first was that the blowout itself didn't supply the spark that ignited the gas. Otherwise, the fire would have started immediately.

Twenty-four seconds is about the length of time it would take a 10-mile-per-hour breeze to waft gas from the pipeline to the campsite. Because of this, Koper suspects that the gas was ignited by something at the campsite—perhaps a campfire or

Coleman stove. Also, that 24-second delay was important under New Mexico law because the length of time during which people are placed in mortal danger affects the punitive damages their survivors can collect. Twenty-four seconds is a long time to be doing the equivalent of staring down the barrel of a loaded gun wondering whether you will be killed. But it's not the same thing as a minute, or five minutes. The ability of the seismic readings to quantify this, Koper says, played an important role in the size of the eventual settlement.

Forensic seismology is a field that is burgeoning so rapidly that it is hard to predict the future. "We're just beginning to eke out every wiggle on the seismogram and what it's telling us," says Wallace. And even though seismometers cost somewhere on the order of $100,000 apiece to install, the number of permanent installations around the globe is only going to increase with each passing decade. "They're always on, and very sensitive," he says, "and it is common for exotic sources to be recorded."

What else these sources might include is up to speculation. Someday soon, one of the new waves of forensically oriented crime shows will stumble onto this technology and base a script around it. All that's needed is to have the otherwise-perfect crime occur close enough to a seismic station for a key event to be recorded, establishing a timeline that shatters the villain's carefully constructed alibi. "Mr. Jones," the brilliant detective could then say, "we know your car hit that abutment at precisely 10:23 p.m., rather than an hour later, as you told the police. That means you ran off the road fleeing the scene of the crime, rather than driving home later on, after spending the evening with your sister, 50 miles away. You were here, crawling out of the ditch, while your sister was trying to establish your alibi by sitting in a pub with someone who looked like you, charging drinks on your credit card. Sisters lie. Seismometers don't."

Forensic seismology is an arena in which reality and mainstream detective fiction might just beat science fiction to the punch.

Author's Note

Forensic seismology was one of those topics I discovered by going to science meetings and randomly sticking my nose into press conferences and symposia. This one is different. I used to live in Corvallis, Oregon, a town that will become relevant late in the chapter. Portions of the story appeared not only in Analog, but in Science and The Sacramento Bee, and it was my editor at the Bee who understood. "This one's personal, isn't it?" he asked.

The answer was yes. And that was before the Lord God bird even entered the story.

HERE BE THERE DRAGONS

Mysteries of an Explored Planet

Once upon a time, when the land was young, there was a magnificent creature, lurking on the fringes of civilization. Maybe it was a roc, maybe a dragon. Perhaps a yeti or a sasquatch or a slavering carnivore carrying away children who ventured too far into the forest. Whatever it was, the elders remember it and pass on the stories. It was real, they say, but none has been sighted in two generations. "It's gone, and you, my child, will never see one." But the child wonders: is it really gone? Or does a remnant lurk on the edges of the civilized world, where few have the time, courage, or inclination to venture?

HERE BE THERE DRAGONS

Speculative fiction is full of such stories. Some are fantasy; some are hard science fiction, set on frontier planets. All draw on the hope that the world isn't fully tamed, that ancient wonders might still linger, if we only knew where to look.

Such hopes aren't unfounded. Change the names a bit, and you have a story that might explode on your evening news any day. No, I'm not talking about Bigfoot. I'm talking about a bird. A bird so magnificent it is sometimes known as the Lord God bird, presumably because the first sight caused old-timers to exclaim, "Lord God, what a bird!" It was a huge, red-white-and-black woodpecker with a 30-inch wingspan, officially called the ivory-billed woodpecker (*Campephilus principalis* for those who like Latin). Once, it lived in old-growth forests throughout the American Southeast. But the last confirmed sighting was in Louisiana in 1944.

For 60 years, experts believed it extinct. Then, in April 2005, ornithologists from Cornell University announced they'd rediscovered it in a section of Arkansas known as the Big Woods. The find was hailed as the ornithological equivalent of finding Elvis alive. It was also a classic case of now-you-see-it, now-you-don't. Within months, another group, led by David Sibley, author and illustrator of the widely used *Sibley Guide to Birds*, challenged the claim. The Cornell team's blurry photos, they said, couldn't be distinguished from the common pileated woodpecker. Maybe the scientists had indeed seen an ivory-bill, but they'd failed to prove it.

Whatever the bird in the photo actually was (and Sibley wasn't the only expert to question it), the debate, conducted in the prestigious pages of the journal *Science*,[14] was enough to ignite the hopes of birders throughout the country. Perhaps, the Lord God bird wasn't extinct after all.

Woodpeckers generally need dead trees in which to peck for insects. That means they do best in old-growth, where there is a good distribution not only of young, healthy trees, but also of the recently dead and dying. The ivory-bill was particularly dependent on mature forests because it fed on the larvae of a large beetle that lived in old, dead trees. "They were specialists in these big beetles," says Greg Butcher, director of bird conservation for the National Audubon Society. Beyond that, the birds weren't terribly picky: at one time they could be found both in upland pines and riparian hardwoods. But gradually, they were pushed into more and more remote areas by the combination of habitat destruction and hunting.

Native Americans had long hunted the bird, says Jerome Jackson, a biology professor at Florida Gulf Coast University and author of *In Search of the Ivory-Billed Woodpecker*, which chronicles his efforts to find the bird in the U.S. and Cuba. The feathers were used for a number of purposes, including decoration on war pipes. "I suspect that the red against black may have signified blood against the hair of an enemy that was successfully scalped," he says. White settlers also killed the birds. "Some people believed the bill was true ivory, so they would shoot them just to get the bill," Jackson says. Others were shot

---

[14] The Cornell find was published on June 3, 2005: John W. Fitzpatrick, et al, *Science* Vol. 308 no. 5727, pp. 1460-1462. Sibley's rebuttal followed on March 17, 2006: David A. Sibley, et al, *Science* Vol. 311 no. 5767, p. 1555. A response by the Cornell team (which stuck to its guns) was published along with the rebuttal.

to be stuffed and mounted or simply by curious people wanting a closer look.

Gradually, the southern forests were converted to cotton fields. But it was only during wartime that logging pressure became intense. "Early in World War I," Jackson says, "Southerners saw war money going to the steel mills of the North. So a bill was passed providing for construction of 1,000 ships of southern pine." Only 320 of the ships were ever built, he says, and none saw service because pine is too leaky for boats. "Most of the wood lay on the ground and rotted," Jackson says. But it was the end of the virgin pine forests of the uplands.

Still, the bird survived. One of its last stands was a 50,000-acre tract of bottomland in Louisiana, which the Singer Sewing Machine Company had bought as a sustainable source of hardwood for sewing-machine cabinets. For a while, after ivory-bills were found there in 1932, the state was able to lease the tract as a wildlife refuge. "But along came World War II, and the wood was needed for pallets to ship [artillery] shells to Europe," says Jackson. "It was needed for coffins. It was 'needed' for all kinds of things associated with the war effort."

The Singer Tract wasn't the only one to suffer this fate. "In the early 1940s, there was apparently a population of about six pairs, nine miles south of Rosedale, Mississippi," Jackson says. "The timber went to the planks of PT boats...One of the worst things for the bird is warfare. We cut timber in our national interest. We 'need' it for this or that and environmental protection usually goes by the wayside."

But logging wasn't the only factor. As far back as the late 1800s, as the birds were becoming rare, hundreds were shot as museum specimens or for private collections. Exactly how big a factor this was in their demise is open to debate. Geoffrey Hill, a biologist at Auburn University, compares it to putting a bounty on the remaining birds. "Deforestation greatly reduced ivory-bill populations," he says, "but collectors took hundreds of birds from the remaining patches...I think it was the shooters that put it out. There's hundreds of these things in museums."

Sibley agrees that the shooting was despicable, but thinks it's an error not to point to habitat destruction as the prime cause. "None of the last known Singer Tract birds were shot," he says. "They just disappeared when the trees were cut down."

Double-Knock on the Choctawhatchee

With the on-again off-again find in the Big Woods, American bird-watchers were a-twitter with rumors. Based on enthusiastic claims in his email, Butcher told me in early 2007, "It seems like [the ivory-bill] is the most common bird in North America."

But unfortunately, subsequent expeditions into the Big Woods haven't found anything new, despite what Butcher describes as the most thorough such search ever conducted (and a reward for a confirmed sighting that eventually reached $50,000).[15] There was even an effort to use a robot birdwatcher. Described at the 2007 meeting of the American Association for the Advancement of Science, this endeavor used two robotic cameras pointing opposite directions across along a powerline cut. The cameras were programmed only to record images of objects moving at 20 to 40 miles per hour—believed to be the speed of an ivory-bill in flight—but failed to find an ivory-bill. Perhaps the failures mean the bird really isn't there. Or perhaps

---

[15] Bob Lalasz, *Talk* (a website of The Nature Conservancy), January 13, 2009.

it's because the Big Woods is indeed big, sprawling over 550,000 acres, a region one-quarter the size of Yellowstone National Park. Even for a large, brightly colored bird, that's a lot of dense timber in which to hide.

But the Big Woods weren't the only focus of attention. In 2005, Hill went kayaking. A decade earlier, he'd gotten a telephone call from someone who thought he'd seen an ivory-bill on Alabama's Pea River but he'd treated it with the skepticism usually reserved for reports of extinct animals. Now, with excitement mounting in the Big Woods, he and two assistants decided to check it out.

The first day was disappointing, revealing little in the way of potential ivory-bill habitat. "We weren't interested in another day on the Pea, so on the spur of the moment we decided to shift," Hill says. So, based on a not-very-detailed map, his team picked the Choctawhatchee, in the Florida Panhandle. "We just kind of blundered into it," he later said. "I didn't even know how to pronounce the name" (It's Chok-ta-HATCH-ee.). Within an hour, one of Hills' assistants spotted what he was sure was an ivory-bill in flight. At about the same time, Hill heard a "double-knock": the unusual pecking sound that distinguishes ivory-bills from other woodpeckers. "It was just to be a weekend outing looking for potential habitat," Hill said. "We never dreamed we'd actually find an ivory-bill."

The following weekend, the team returned to the river, where another of Hill's assistants reported a clear view of a female ivory-bill. The assistant didn't have the opportunity to snap a photo, but said he saw the distinctive plumage of the ivory-bill, distinguishable from a pileated woodpecker by a white trailing edge on the upper wing, white stripes down the back, and an all-black crest.

Hill and his colleagues are confident of their discovery but aware they have yet to prove it. "The only evidence that would constitute irrefutable proof is a clear photograph or video," he says, "and such an image has to date eluded us." His team, however, recorded 14 sightings and worked with Canadian scientists to collect 10,000 hours worth of sound recordings from unmanned listening stations, tallying 300 sounds matching those believed to be made by ivory-bills. They also found nest cavities too large for other local birds and places where very big woodpeckers appear to have been pecking on trees, publishing their findings in 2006, in the journal *Avian Conservation & Ecology*.

Still, without photos, everyone else was cautious. "Nothing is confirmed, but there is a lot of good evidence," the Audubon Society's Butcher said. "They seem to have found some very good habitat and have been very diligent in trying to document it." Sibley, who helped deflate the Big Woods claim, was even less optimistic. The findings, he told me, were "intriguing," but he noted that the Choctawhatchee region is small, with relatively young trees, and is heavily traveled by boaters, fishermen, and hunters. "I think the sighting and audio evidence is questionable and mostly based on wishful thinking," he said. The hunt, however, continues. "If the elusive ivory-billed woodpecker exists, the Choctawhatchee River basin in Walton County is the perfect habitat to explore the possibilities," an organization called Walton Outdoors posted on its website as recently as 2011.

Corvallis Resurrection

Birds can be hard to photograph. Not so with other species.

HERE BE THERE DRAGONS

In the spring of 1999, Rana Foster was volunteering for the National Audubon Society, monitoring bluebird nests, when she spied an unusual flower in a preserve in Corvallis, Oregon. The plant was only an inch tall, but bedecked with striking blossoms like purple trumpets streaked with white and yellow. It was growing in a muddy channel where a few months earlier a river had jumped its banks in an unusually high spring flood.

Foster described her find to Steve Northway, an amateur botanist helping the city restore the preserve from four decades of rye grass cultivation. The moment he saw the flower, Northway knew what it was: a type of wild snapdragon called the vernal pool monkeyflower, or *Mimulus tricolor*—once plentiful, but thought to have been extinct in Oregon since 1991.

In the next few days, Northway tallied a thousand more *Mimulus* plants in the 73-acre preserve, guessing he'd only found about a third of them before he tired of counting. A month later, I was sprawled on the ground before him and Foster, examining one of the flowers close-up through a camera lens. I had the odd sense that neither it nor I was supposed to be there—that it might vanish if I reached out to touch its fuzzy leaves. For the past hour, Northway and Foster had been discussing the plant's startling return. Northway speculated that the seeds' hard, nut-like capsules had survived for years, buried in the sod until the flood stripped the grass away and gave them their chance.

The *Mimulus*, he explained, had once thrived both in flood channels and vernal pools: ponds that lasted late into spring, then dried to caked mud during the Willamette Valley's four-month summer drought. The rapid shift from puddle to near-desert made it hard for other species to grow, creating a niche where the ground-hugging monkeyflower could germinate, bloom, and die in a matter of weeks. Only a few decades ago, the flower had speckled the 100-mile-long valley with splashes of color. But plowing and stream diversions had destroyed so much of its habitat that eight years previously, it had disappeared from Oregon—although a remnant population still clung to life in California. And now, amazingly, it was back: a miracle in a city park.

If the ivory-billed woodpecker has survived, it's obviously not because eggs have slumbered through the decades. Rather, a few survivors must have lingered, undetected, for more than 60 years.

That's not impossible. The California sea otter was believed to have been driven to extinction early in the Twentieth Century, killed off by fur hunters seeking to make their fortunes from its incredibly fine pelt. But a tiny colony, perhaps only a few dozen, survived off the rugged Big Sur coast, protected by ranchers who refused to tell the outside world of its existence. Then in 1938, word got out and a photographer confirmed it with a grainy black-and-white image of some 75 otters—quite possibly the entire population—in a single, large group. Today, Pebble Beach golfers can hardly fail to spot the creatures, only a stone's throw offshore.

Nobody appears to have been hiding a colony of ivory-bills. But for half a century, their forests have been re-growing. If a few managed to survive. . .somewhere, somehow. . .then maybe, just maybe, they're starting to reclaim their one-time range. "There's every reason to believe that a population could expand if there are still breeding pairs around," Butcher says.

One place where the ivory-bill might have survived is along

Florida's Suwannee River. "[The Suwannee] is the heart of where the birds were once abundant," Jackson says. "That really was the homeland for ivory-bills. Where the Suwannee goes into the Gulf is really wild." There were once so many woodpeckers there, he says, that about half of the 400 specimens now in museums were collected in the region.

If the bird does survive, Jackson thinks it's been protected in part by the modern hunting culture of the South. "Large [private] forest areas have been preserved as places to hunt," he says. "They go in and hunt deer in the fall and turkeys in the spring, and the rest of the time it's left alone. Those are places where the ivory-bill might still exist." Jackson himself made a determined effort to find the bird along the Suwannee in the late 1980s. He's convinced it persists, but, like other ivory-bill searchers, failed to find definitive proof.

Today, the river is being promoted as a recreational region, but Jackson doesn't think that will deter ivory-bills from repopulating the area. "I don't think people in kayaks or canoes are going to make a lot of difference," he says. Butcher agrees. What the bird really needs, he says, is several square miles of undisturbed habitat. "They're a big bird, so they need a lot of food, and since the food they eat is fairly specialized, they need a big area in which to find it. But there's no reason to suspect that they'd be more disturbed by hunters, boaters, or birdwatchers than any other bird," he says.

A bigger threat is real-estate development. "If we have any hope for ivory-bills in the area, we've got to find them now and get the habitat protected," Jackson says. "It's not going to be there in 20 years. It might not be there in ten years."

RICHARD A. LOVETT
Disappearing *Delphinium*

Although it was June, my day in Corvallis had been darkly cloudy, as a dreary spring refused to heed the dictates of the calendar. But eventually, sunlight burst between late-afternoon clouds. I put away my camera and stood up, observing a prairie transformed to an eye-searing expanse of green.

Northway asked if I wanted to see another flower— desperately endangered, but not yet extinct: the peacock larkspur, or *Delphinium pavonaceum*. It, too, was once common, but existed now in only a handful of stands. "Come see, so that when we're old there will be a few of us to testify that it really existed," he said. It was an impossible invitation to pass up, so Northway, Foster, and I caravanned along a succession of farm roads until he pulled to the shoulder. "There," he said, indicating a plant like a tall, ragged lupine. Most larkspur are deep blue with light-colored centers, but this one has reversed the colors, with white flowers boasting pale lavender centers.

"Ten years ago, we couldn't have driven out here without seeing these everywhere," he said. But now, the state had tabulated only 53 patches, of which half were probably already gone: killed off by unknowing swipes of hay-balers or mowers. The rest were all that remained in the entire world.

Northway had been monitoring five of the known patches. This particular one had been thriving. "Four years ago, there were only 70 plants," he said. "This spring I counted 476." He had concluded that there was no sense in hiding such discoveries. Experience had taught him that the biggest risk was an inadvertent mowing, plowing, or herbicide spraying. Like the Big Sur ranchers protecting the sea otter, most farmers are thrilled to give up a single bale's worth of hayfield to help spare a remnant of a shrinking past.

Northway left, but I lingered as the last of the sunlight graced his elegant *Delphinium*. It was the summer solstice—the longest day of the year—and somehow that seemed appropriate. It's both a day of jubilation and sorrow, simultaneously marking the onset of summer and the fact that now, the days would grow increasingly shorter as the world spiraled toward winter. I wondered which lay ahead for the *Delphinium*: a summer renaissance, or a long slide toward oblivion. Only three days earlier, Northway's determination not to hide his endangered species had been put to a severe test when somebody destroyed one of his other four patches. The flowers hadn't just been picked by someone who'd found them pretty. They hadn't been mowed by accident. They'd been selectively pulled up by the roots in what could only have been an act of malice. "There were five [patches]," Northway said softly, shortly before he left. "Now there are four." It would take so little to save this plant, but the oncoming winter loomed closer than the warm summer twilight.

Northway also told me that the flower, rich with nectar, was once the primary early summer forage for Willamette Valley bumblebees, now themselves in decline. As I watched the *Delphinium* sway in a breeze blowing in from the distant Pacific, I couldn't remember the last time I'd seen a bumblebee.

It is easy for stories like this to become depressing. Progress marches; *Delphinium* and woodpeckers fall by the wayside. But it isn't always that way. Sea otters and *Mimulus* miraculously reappear. That makes this a story of anticipation, and still-open frontiers.

RICHARD A. LOVETT

More than anything else, I am a wilderness enthusiast. The week before starting the first version of this story, which has appeared in various forms in *Science*, *The Sacramento Bee*, and *Analog Science Fiction and Fact*, I spent four days in the Oregon Cascades, climbing a 10,000-foot volcano and scrambling over glacial moraines to sit for hours above turquoise, green, and azure lakes too tiny to appear on my map. I thought my trip was about solitude and relaxation. Now, I'm not so sure.

Until an ivory-bill is definitively sighted, nobody will know whether the species has survived. But the recent sightings are more than just tantalizing. "The reports in Arkansas and Florida give us hope," says Jackson. And while hope is not proof, he says, "it is the fire that incites us to seek the truth." Amateur birders agree. "When I first heard about the sightings in Arkansas, I was skeptical," says David Hatfield, a birder from Portland, Oregon. But later, he changed his mind. "That such a large and colorful bird appears to have survived is good news for the ivory-bill, a positive note on the state of our wildernesses, and a fantastic story," he says.

In ancient times, I was once taught, maps had boundaries, beyond which lay unknown terrain. Sometimes the unknowns were simply blank. Other times they were stamped with a phrase that in my youth I took as a warning: Here be there dragons. Now, I see those dragons in a different light. Maybe they're resurrected wildflowers. Maybe they're ivory-billed woodpeckers. Maybe they really are the things we like to write about in science fiction. But above all, what they are is hope.[16]

---

[16] A different kind of hope also persists for the *Mimulus* tricolor and the peacock larkspur. Seeds from both plants are being cultivated by the Center for Plant Conservation's Berry Botanical Garden in Portland, Oregon.

Author's Note

From monkeyflowers and woodpeckers, let's segue back to space. The next story started out in Cosmos magazine (a gorgeously produced Australian science magazine that deserves wider circulation). The science stems from NASA's Cassini mission, which has been orbiting Saturn since 2004. Unfortunately, its maneuvering fuel is running low, and its explorations won't continue beyond 2017. Meanwhile, it's given us 13 years of spectacular data.

DANCE OF THE RINGS

When Galileo first looked at Saturn in 1610, he saw a planet with a blurry brim he interpreted as two large moons, one on each side. But two years later, the "moons" had vanished. . . only to reappear two years after that. Galileo was puzzled. The best explanation he could come up with was that Saturn had some sort of wings or arms, sometimes spread wide, sometimes tightly folded.

Today, we know he was seeing the rings and that they disappeared because Saturn passed through a phase in which they were nearly edge on from Earth, invisible to his crude telescope. But 400 years later, there are still plenty of mysteries about the rings—even though the high-resolution cameras of the Cassini spacecraft, which has been orbiting Saturn since 2004, have now photographed them from nearly every conceivable angle.

Not that we haven't learned a lot. We know, for example, that the rings are extremely thin—only about five to ten meters thick. "That's like a two- or three-floor building," says Phil

Nicholson, a planetary scientist from Cornell University in Ithaca, New York. Furthermore, from one side of Saturn to the other, the rings are almost 300,000 kilometers wide. "That's a ratio of 30 million to one," Nicholson says. "It makes this sheet of paper look pretty thick in comparison." So thick, in fact, that if paper actually had the same relative thickness as the rings, you could pile thousands of copies of this book in a one-inch stack. "The rings," Nicholson says, "are the most two-dimensional structure we know in the Universe."

But they are also bright enough to easily see and dense enough to block starlight. That means the ring particles, which range from about the size of golf balls to refrigerators, must be very tightly packed. "They're only a shoulder-width apart," says Nicholson. "They're having collisions all the time." These collisions damp out differences in their motions, causing nearby particles to orbit at nearly the same speed, with velocity differences of only a few millimeters per second. "Imagine a bunch of horses running around a racetrack at almost the same speed," Nicholson says. "One jockey can reach out and pass his whip to another even though the horses are going 35 to 40 miles per hour."

Officially there are seven rings, known as the A, B, C, D, E, F, and G rings. These names might be the only simple things about the ring system—and even they aren't as straightforward as they sound. They're named in order of discovery (approximately the same as their order of brightness), which is not the same as order from Saturn. The innermost ring is the D ring. From there, it goes C, B, A, F, G, E. And the G is so faint it wasn't even seen on the earliest Saturn fly-by, by the Pioneer 11 spacecraft in 1979. Rather, it took the better cameras of

Voyager 1 to discover it a year later. But these are merely the official divisions. Cassini's best photos reveal hundreds, even thousands, of ringlets within each major one.

There are also countless other structures. The closer we look, in fact, the more complex the rings become. Scientists have known for some time, for example, that small moons a few kilometers in diameter, orbiting within them, can gravitationally

NASA/JPL/Space Science Institute

67

sweep ring particles aside, creating gaps such as the A ring's 325-kilometer-wide Encke Gap, patrolled by the 28-kilometer moon, Pan. Other gaps, such as the enormous Cassini Division (4,800 kilometers wide, separating the A ring from the B), are created by the gravitational effects of more distant moons. But closer examination reveals these gaps to have wondrously filigreed edges, sculpted by even more complex interactions with the moon's gravity.

There are also radial stripes, called spokes, first seen by Voyager in 1980, but—in an odd echo of Galileo's vanishing "arms"—not seen again until 2007, even though Cassini had been looking for them for three years.

Scientists now think that, like Galileo's "arms," the spokes also might be linked to Saturn's 29.7-year orbit. But in this case, it's not the tilt of the rings as viewed from Earth that makes them come and go. Rather, it's the slow parade of Saturn's seasons. The hypothesis is that the spokes are formed by microscopic particles suspended above the rings by static electricity—bright in some lighting conditions, dark in others. The static electricity, this theory holds, comes from the interaction of Saturn's giant thunderstorms with its magnetic field, and the come-and-go nature of the spokes simply represents Saturn's changing weather as it moved toward its equinoxes, one in 1980, and another in 2009.[17] (There would have been another equinox in 1994, but there was no spacecraft there at the time to take pictures.)

Also included in Saturn's ring zoo are "propellers," so named

---

[17] Jones G. H., *et al*, *Geophysical Research. Letters*, 33, L21202 (9 November 2006).

because they look like old-fashioned airplane props, only a few kilometers wide, and tens of thousands of kilometers long. Some of the largest have even been unofficially named for famous aviators. History buffs will recognize Bleriot, Earhart, and Santos-Dumont, among others, says Carolyn Porco, head of the Cassini imaging team.

Porco believes the propellers are created by moonlets—or even wannabe moonlets called "clumps"—too small to open full, 360-degree gaps. Instead, they open partial gaps, which eventually fill in, both in front and behind, around the curve of the rings. Not surprisingly, they lie in the densest part of the rings, where clump formation is gravitationally easiest. The biggest are in the outer sections of the A ring, but enough have been found that there is an unofficial "propeller zone" elsewhere in the A ring, where hundreds of smaller propellers mark even smaller moonlets.

In the years since 2005, when the first propeller was seen, scientists have been able to track them, watching how their movements change in the rings' complex gravity. That, Porco says, allows them to track moons too small to see (perhaps one kilometer in diameter), letting them use the rings as a laboratory for watching, on a smaller scale, the evolution of the early Solar System. "Scientists have never tracked disk-embedded objects anywhere in the universe before," says Matthew Tiscareno, also of Cornell University. "All the moons and planets we knew about before orbit in empty space."

"It allows us a glimpse into how the Solar System ended up looking the way it does," adds Porco.

Yet another intrigue was discovered in 2009, when Saturn

passed directly through its equinox. At that time, the rings were almost perfectly edge-on to the Sun, and for a brief interlude, never-before-seen vertical features rising above the plane of the rings were thrown into sharp relief, much as the setting sun on Earth casts mammoth shadows from every tree, every blade of grass.

The biggest (as in tallest) of these were along the outer edge of the B ring, which turned out to have peaks rising as high as 3.5 kilometers above the ring plane. It was a totally unexpected find, but in a study in the November 1, 2010 issue of the *Astrophysical Journal,* a team led by Joseph Spitale of the Space Science Institute in Boulder, Colorado believed they found the answer.[18] That part of the rings, the scientists concluded, probably contains an entire population of moonlets that migrated through the ring to its edge then became trapped by a gravitational resonance with the moon Mimas, which lies 67,000 kilometers farther out.

Such resonances (also discussed in Chapter 1) occur where the relative orbital positions of two objects (in this case, ring particles and a moon) repeat, orbit after orbit, allowing gentle tugs to cumulate until they have warped the rings out of their natural shape. At the same time, Spitale's team found, the rings can produce their own oscillations, somewhat like water sloshing in a bathtub. These—just like the water in the tub, reflect off the sides (in this case the gap at the outer edge of the B ring) producing startlingly large sloshes. Or, if you prefer a musical analogy, it's like plucking a guitar string. Like the guitar, Spitale says, "The ring has its own natural oscillation frequencies. That's

---

[18] Spitale, & Porco, *The Astronomical Journal,* 140(6), p. 1747 (2010).

what we're observing."

Vertical oscillations can also be studied by watching changes in the light of stars passing behind the rings. "A reasonable analogy is a person walking behind a picket fence, watching the light from the setting sun come through," Nicholson says. In this manner, he has found a narrow gap in the C ring, with a raised edge on one side and a dipped one on the other, each about 1.5 kilometers high. "It's like a snapped-up hat rim, Humphrey Bogart style," he says. This feature and adjacent ripples seem to have been set off by a resonance with the gravity of the giant moon, Titan. "It's yanking the ring particles up due to the inclination of its orbit," he says.

The waves, he adds, propagate by gravity, as rising particles first tug up their neighbors, then are pulled back into the plane of the rings by the overall gravity field. He compares the result to a slow-motion tsunami. "The amplitude is about 500 meters," he says, noting that this would be huge for a real tsunami. "But it's moving pretty slowly, about 250 meters a day."

Other wave-forming processes are even more exotic. In a 2010 study of Jupiter's rings (far less prominent than those of Saturn), Mark Showalter of the SETI Institute in Mountain View, California, and Joseph Burns of Cornell University concluded that mysterious ripples in these rings, first observed when the Galileo spacecraft passed by in 1996, were caused by gravitational disruptions from something passing through the rings en route to an impact with Jupiter. Since then, the two scientists reported at an October 2010 meeting of the American Astronomical Society's Division of Planetary Sciences in

Pasadena, California that the ripples have been winding tighter and tighter, like coiling rope. By mathematically turning back the clock, they uncovered not one, but two sets of ripples, one tracing to 1994, and the other to 1990.

The simple passage of an asteroid through the rings, like a giant bullet, couldn't have produced such widespread disruption,

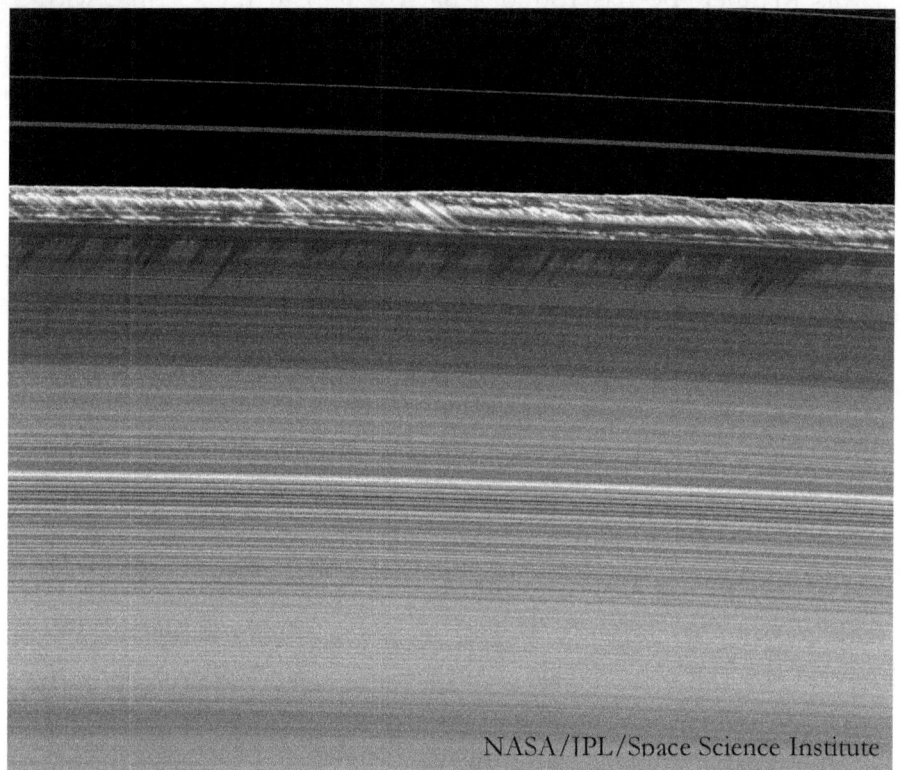

NASA/JPL/Space Science Institute

they added. Rather, it would have required a shotgun blast of smaller particles, such as would be produced by a broken-up comet. In other words, one set of ripples was created by the impact of comet Shoemaker-Levy 9 into Jupiter in 1994,[19] and

---

[19] See Mark R. Showalter, et al., *Science*, 6 May 2011, pp. 711-713.

the other by an unknown impact four years earlier. And on Saturn? Similar ripples, they said, have been discovered in the planet's C and D rings, where they appear to reflect an event in 1983 or 1984. "The rings are witness plates for cometary impacts, giving their frequencies and sizes," Burns says.

But the rings' greatest intrigue is their most fundamental. Where do they come from? They contain enough mass to form a moon about 400 kilometers across, so, not surprisingly, traditional theory says that's how they started: as a 400-kilometer moon that strayed too close and was torn apart by Saturn's mammoth gravity. There's just one problem. The rings are made almost entirely of ice—90 to 95 percent pure, says Robin Canup, a planetary scientist from the Southwest Research Institute in Boulder, Colorado. But the primordial Solar System is believed to have been composed of about equal parts rock and ice. So, if the rings are formed from a smashed-up moon, what happened to the rock?

Some scientists have attempted to get around this by positing that the rings' progenitor was a giant comet, rather than a moon. That would have meant a higher fraction of ice, perhaps, but Canup doesn't think it's enough. "It's not clear that that would give you a pure ice ring, either," she says. And, she notes, Saturn is the *least* likely of the giant planets to have had such a comet-bursting encounter. Neptune and Uranus lie closer to distant zones where comets originate, and Jupiter's even-larger gravity more easily draws such bodies into it, such as Shoemaker-Levy 9. "[Such an encoun-ter] is ten times more likely at Jupiter, Uranus, and Neptune than at Saturn," Canup says.

Instead, she thinks the rings came not from the demise of a

small moon, but a large one—perhaps the size of its giant moon Titan (5,100 kilometers in diameter). In a paper in the December 16, 2010 issue of *Nature*,[20] she suggests that such a moon could have spiraled into the infant Saturn, shedding an icy mantle while its rocky core, made of tougher stuff, remained intact until its final plunge into the planet. "If the core hits the planet's surface before it disrupts, the end result is a pure ice ring," she says.

"It's a very clever new idea," adds Burns. "One of the things it can do is produce rings made out of quite pure water ice, which has been a problem in the past." Furthermore, he says, it makes sense that the outer shell would be fairly pure water, because the doomed moon would have heated up from tidal friction as it spiraled inward. That would have softened it enough, he says, for the "heavy stuff" to settle to the center. Furthermore, computer models show that large moons, if they form early in the evolution of a gas/dust cloud such as that believed to orbit the infant Saturn (or the early Sun, for that matter) easily migrate inward due to gravitational interactions with remaining gas in the inner parts of the disk. (A similar form of gas drag was discussed in Chapter 1, regarding protoplanets). It's a process called Type 1 migration and its speed is strongly dependent on the moon's size. "We think that when they reach a mass comparable to Titan, they start to migrate," Canup says.

The mantle of a Titan-sized moon would contain a lot of ice; about 1,000 times more than presently remains in the rings, Canup says. But as the ice spread, much would have fallen into Saturn. Some would also have condensed into new moons—and in a study in the June 9, 2010 issue of *Nature*, a team led by

---

[20] Robin M Canup, *Nature*, 16 December 2010, pp. 943–926.

Sébastien Charnoz of Paris Diderot University, France found that gravitational interactions would have the opposite effect on *these* moons as on the giant moon that produced them. Rather than spiraling into Saturn, Canup says, these "spawned moons" would migrate outward, as though recoiling from their birthplaces.[21]

That solves yet another riddle. These "spawned moons" would include Enceladus, Dione, and Tethys, all of which have densities low enough that they must be composed primarily of ice. Scientists had been puzzled by how such icy moons could form, but Canup's theory answers the question, almost by accident. "It occurs very naturally as part of another origin process: the process by which the rings themselves were made," Burns says. "She has a pretty convincing story, I would say."

Furthermore, there's no reason this entire process had to have happened only once. Currently, Saturn has one large moon. Jupiter has four. But perhaps Saturn, like Jupiter, once started with several, each spiraling toward the same fate: one after the other, until eventually the gas cloud dissipated. Under Canup's hypothesis, it's possible to imagine a chain of Titan-sized moons marching to their deaths, until only one remained when the system stabilized—with the present rings formed by the last big one to die.

Ultimately, however, the rings' greatest contribution to science is as a close-at-hand laboratory for studying other astronomical processes.

---

[21] See Sébastien Charnoz et al., *Nature* 752-754 (2010) *and* Charnoz, et al, *Icarus 216(2),* 535-550 (2011).

Spatale's team compares them to a spiral galaxy. Nicholson suggests a protoplanetary disk. "The theoreticians assume there are gaps in these disks as well," he says, referring to his Humphrey-Bogart hat-brim "crack" in the C ring. "Understanding the physics is important," he says. "Saturn's rings are probably the laboratory for studying this."

The next step, he hopes, is a not-yet-formally-proposed mission called the Saturn Ring Observer, in which a spacecraft would be placed in an orbit close to the rings—perhaps only 3 to 5 kilometers above them, so that it would travel with the ring particles "a bit like the pickup that precedes harness-track horses. "The craft would slowly drift across the rings, Nicholson says, "like a laser playing a CD, building up a map of the entire system over several months."

"To some of us," he adds, "this is the ultimate in cool."

Author's Note

My undergraduate university didn't offer minors. Had it done so, I might have been the only one of 40,000 students to graduate with a major in astrophysics and minor in history. The next four stories started out by piquing my interest for the history, then expanded…

THE DAY THE SUN EXPLODED

It began on a late-summer morning in 1859, when British astronomer Richard Carrington was observing the Sun by projecting its image onto a glass plate and meticulously sketching what he saw. The son of a wealthy brewer, Carrington had invested his inheritance in a private observatory south of London, where he'd spent years studying the Sun. Already he had two major finds to his credit. One was the Sun's differential rotation: the rather startling discovery that it does not rotate as a solid body but instead spins more quickly at the equator than at the poles. The other was what is now known as the "butterfly effect," in which the zones where sunspots form move ever closer to the solar equator as the sunspot cycle proceeds—like a butterfly slowly flapping its wings.

But on the morning of September 1, Carrington saw something unprecedented: a pair of bright crescent shapes forming near a large group of sunspots. At first, he thought a stray shaft of light had gotten into his instruments. But a quick check revealed that whatever he was seeing really was happening on the Sun.

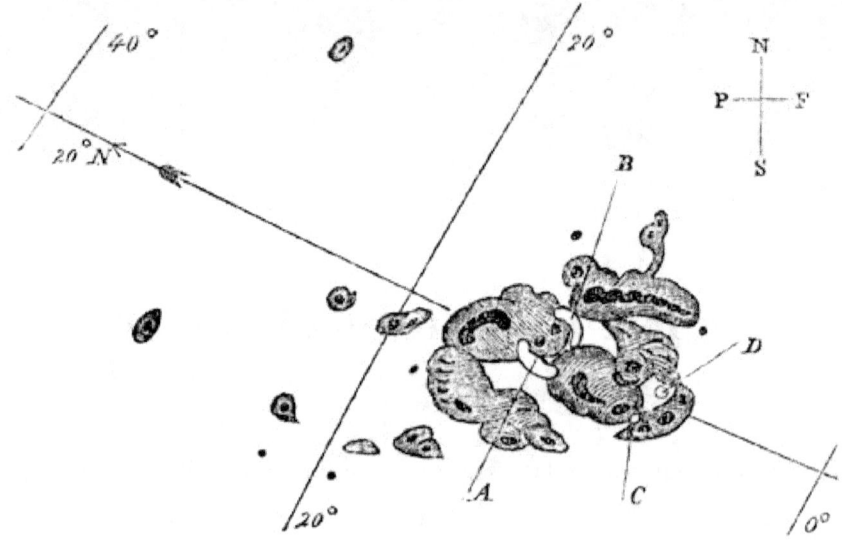

Carrington's original 1859 drawing

Today, Carrington would have been frantically snapping photos (or, more likely, recording everything on a computer), but in 1859, what he needed was a witness. Excitedly (or in his own words, "somewhat flurried"), he dashed through the observatory offices, trying to find someone, anyone, to confirm what he'd seen. But the building was deserted. Frustrated, he ran back to his telescope, where he was horrified to find that in the mere minute he'd been gone, the bright arcs had already faded. Shortly afterward, they vanished completely.[22] "It lasted all of

---

[22] Luckily, another astronomer, Richard Hodgson, happened to be looking at the Sun at the same time and confirmed Carrington's discovery.

five minutes," says Ed Cliver, a space physicist and Carrington historian at the U.S. Air Force Research Laboratory in Bedford, Massachusetts.

What Carrington had seen was a rare white-light flare, herald of a particularly violent eruption of hot plasma from the Sun's surface. During active periods, the Sun can produce thousands of flares a year, but according to a tally by Cliver, less than 0.1 percent are white-light. And even among these, Carrington's must have been huge. Large enough that if it happened again today it would be a disaster like nothing the modern world has ever seen.

As of the time this book went to press, the Sun was reaching its latest activity peak with no truly major solar events. The last major solar storms to hit the Earth were in late October and early November of 2003, when a series of giant solar eruptions that came to be known as the "Halloween storms" hurled enough energy at us to interfere with satellite communications, produce a brief power outage in Sweden, and light up the Halloween skies with ghostly auroras as far south as Florida and Texas. But compared to Carrington's "event," these were a mere "pipsqueak," Cliver's Air Force colleague, space physicist Peggy Ann Shea, said shortly after the Halloween storms, at a joint meeting of American and Canadian geophysicists.

Within minutes of Carrington's sighting of the flare, the magnetometers at Britain's Greenwich and Kew observatories gave large jerks as something radically altered the Earth's magnetic field. Then, 17½ hours later, the Earth's magnetic field truly convulsed. This time, the magnetometers at Greenwich and Kew jumped off scale and stayed there for hours. Northern lights spread as far south as Cuba and Honolulu. (Southern lights were seen as far north as Santiago, Chile.) "People in the Northeastern U.S. could read newspaper print just from the light of the aurora," says space weather expert Daniel Baker of the University of Colorado's Laboratory of Atmospheric and Space Physics. But the sky show wasn't the only excitement. Telegraph operators reported sparks leaping from their equipment, melting wires and starting fires. "That is an indication the storm was truly exceptional," said Bruce Tsurutani, a space-weather expert at NASA's Jet Propulsion Laboratory in Pasadena, California.

At the time, nobody really understood what had happened. Carrington (and others) clearly noted the correlation between the flare and the geomagnetic storm. But it was the first flare ever seen, and the scientists of that era were a particularly cautious lot. "While the contemporary occurrence may deserve noting, [Mr. Carrington] would not have it supposed that he even leans toward hastily con-necting them," Carrington wrote in an 1860 paper published in the *Monthly Notices of the Royal Astronomical Society*. Then, quoting Aristotle, he added, "One swallow does not make a summer." But reading between the lines, Cliver believes Carrington knew it had been no coincidence. That carefully worded statement, he says, "has at least one too many qualifiers in it."

Tangled Forces

The physics of geomagnetic storms is complex, but the basic ideas are simple. Storms begin deep in the Sun's interior, where swirling plasmas snag the Sun's magnetic field, coiling it into ever-more-convoluted shapes. Every now and then, the tortured field springs into lower-energy configurations, like a mousetrap snapping shut. In the process, it releases enormous amounts of pent-up energy. Even the 11-year cycle is related to this: as it progresses, the Sun's magnetic field becomes increasingly turbulent until eventually it releases enough energy in a flurry of solar storms to return to a more relaxed state…only to start the process anew. Sunspots mark locations of particularly strong turbulence, and flares mark the strongest outbursts. Sometimes they are aligned in the right direction to hurl their power in our direction.

When outbursts occur, the energy can come in two forms. The first is electromagnetic: not just visible light, but X-rays and ultraviolet light that hit instantaneously with our first sighting of a flare. This radiation is strong enough to ionize gas molecules, temporarily altering the Earth's magnetosphere (a tenuous layer extending several times the Earth's diameter into space). During the Carrington Event it was this that caused the first round of disturbances at the Kew and Cambridge magnetometers.

But that's just the beginning. A storm can also be accompanied by a coronal mass ejection (CME), in which ionized gas is blasted out at millions of kilometers per hour. This plasma cloud carries the bulk of the storm's energy, and when it hits the Earth's magnetosphere it alters it even more dramatically than the X-rays and ultraviolet light before it. (It also poses a radiation danger to astronauts.) The result: brilliant auroras and magnetic surges so strong that any long metal object on Earth (such as a telegraph wire, electrical transmission line, or even an oil pipeline) can act as a giant antenna, picking up energy from space.

So, just how big was the Carrington Event?

Until recently, space physicists thought the worst the Sun could throw at us was a storm that struck on March 13, 1989. That one produced once-in-a-lifetime auroras extending nearly to the tropics and sent electrical surges ricocheting through the power grid of Canada's Hydro-Quebec. Those surges were strong enough that within seconds, they knocked out power relays like a chain of dominos, leaving the 6 million residents of Quebec without power for nine hours on a winter night. Then, scientists started taking a new look at Carrington. And in a quest that ranged from Mumbai to Greenland, they kept coming up with the same answer—cowabunga: the Carrington Event must have been huge. Large enough that in comparison, the great Quebec Blackout Storm was a mere kitten.

"Imagine large cities without power for a week, a month, or a year," says Baker, lead author of a report on the topic from the U.S. National Research Council.[23] If it happened today, he added at a 2010 meeting of the American Geophysical Union, "the losses [in the U.S. alone] could be \$1 to \$2 trillion, and the effects could be felt in the economy for years."

Tracks of the Monster

The first sign that we might have to substantially upgrade our worst-case scenario came in a 2003 paper by Tsurutani, in the *Journal of Geophysical Research*.[24] In that paper, Tsurutani reported

---

[23] Severe Space Weather Events: Understanding Societal and Economic Impacts, The National Academies Press (2009).
[24] Tsurutani, et al, *J. Geophys. Res.*, 108(A7), 1268 (2003.

that although the Kew and Cambridge magnetometers provided no meaningful data during the main portion of the storm, a long-forgotten instrument near Mumbai, India continued to give readings. That instrument was a bit old-fashioned: instead of using a stylus to automatically etch its readings onto a moving roll of paper, it needed a human attendant to write them into a logbook. But low-tech methods have their advantages. When this instrument's readings also started to go off-scale, the attendant simply turned down the gain and continued taking notes.

Based on a model built with data from more recent solar storms, Tsurutani used these records to calculate that at peak the Carrington Event caused the Earth's magnetic field to fluctuate by 1760 nanotesla (nT)—roughly a 4% change. To put that in perspective, the largest such fluctuation ever recorded by modern instruments was 590 nT, during the 1989 storm. In other words, according to Tsurutani's calculations the Carrington Event kicked off a geomagnetic storm three times more intense than the one that knocked out Quebec's power. There has been some dispute among geophysicists over whether Tsurutani may have over-compensated for the changed instrument settings, but there has been little dispute that these readings probably reveal the most intense magnetic storm ever known.

More evidence of the storm's power comes from the speed with which the CME reached Earth. After a typical flare, the delay is about 30 hours—meaning the particles are traveling at about 5 million kilometers per hour, or 3½ times faster than the normal solar wind. At 17½ hours, the Carrington Event is the second-fastest on record. Top honors go to a storm in August

1972, in which the shock wave was only 14½ hours behind the flare. (The CME from the Halloween storms of 2003 took 19 hours to hit Earth.)

On first consideration, it might seem reassuring that on this scale, the Carrington Event ranks only second. Could it mean that the Mumbai magnetometer operator hadn't turned down the gain by as much as Tsurutani thought, causing his century-and-a-half-later calculations to be in error? Probably not. Solar storms are complex creatures, composed of many factors. Compared to Carrington's, in fact, the 1972 storm was another kitten. According to a website that lists major events from 1859 to the present,[25] the 1972 storm's northern aurora extended only as far south as Illinois and Colorado, and, while that storm was strong enough to interfere with communications (to the extent that taxi drivers reportedly found themselves getting calls from thousands of kilometers away), the overall effects were minor. Despite the speed of its CME, this storm was a long way from being Carrington redux.

The reason, Tsurutani says, is that geomagnetic storms are affected not merely by the speed of the ejected plasma, but by the quantity of charged particles hitting the Earth and the manner in which the magnetic fields carried by them interact with the Earth's own. In the Carrington Event these magnetic forces might have interacted in just the right (or wrong) way to produce a superstorm. But it's also possible that its CME carried vastly more solar particles than the 1972 one.

To test the latter hypothesis, a team of researchers that included Shea and her husband, Donald Smart, turned to ice

---

[25] www.solarstorms.org.

cores from Greenland's Summit Station (which, as its name suggests, sits at the crest of the Greenland ice sheet). When high-energy particles from a CME hit the upper atmosphere, they trigger a cascade of chemical reactions that, among other things, produce nitrates. These settle to Earth, where they are preserved in polar snows—and, ultimately, Greenlandic ice. It's a unique form of solar-wind pollution that serves (for those with sensitive enough instruments) as a marker for big flares. Better yet for Shea and Smart's investigations, these nitrates are most strongly produced by fast-moving protons with energies greater than 30 million electron-volts (MeV) —precisely the super-energetic particles that pose the greatest radiation risk to space equipment and astronauts, as well as creating the magnetic disturbances that light up the night skies and ruin power equipment on Earth.

Using the ice cores, Shea and Smart found nearly a score of giant storms between 1561 and 1992. And, no surprise, Carrington was the largest. In fact, Baker says, by this measure "the Carrington Event stands head-and-shoulders above other events—probably by a factor of about four larger than anything in the modern space era"—including the 1989 and 1972 storms.

Biggest on Many Scales

The Carrington Event was impressive on many levels, but Cliver thinks that one of the most impressive is simply that it stands at or near the top in every measure used to determine its power: time-of-flight delay between the flare and the arrival of the CME, the number of 30-MeV particles, the strength of the magnetic disturbances, the latitudes at which aurora were visible. "It's exceptional in each category, and I can't think of any other event for which I can say that," he says.

In 1859, psychedelic skies and melting telegraph wires weren't a big deal. Aurora did no harm, and telegraph wires could also be taken down by windstorms. But what would it be like if something similar happened today?

Not good is the scientific consensus.

To begin with, says Baker, electrical disturbances like those that took down parts of the telegraph system—"the Internet of the era"—would be far more disruptive to modern electronics. "We live in a cyber cocoon enveloping the Earth," he says. "Imagine what the consequences might be."

We're not talking just about your electric garage door opener or cell phone, which might go wonky from the type of errant signals that in 1972 had cab drivers fielding dispatch calls from the far side of the country. "What's at stake," says Tom Bogdan, director of the U.S. National Weather Service's Space Weather Prediction Center in Boulder, Colorado, "are the advanced technologies that underlie virtually every aspect of our lives."

The National Oceanic and Atmospheric Administration (or NOAA, which is the parent organization of the National Weather Service) breaks solar storms into three categories: radio-blackout, radiation, and geomagnetic. Not that it's a hard-and-fast taxonomy: some storms fall into all three categories. Each category is further rated on a five-point scale, ranging from "minor" to "extreme," allowing a total of 15 ratings.[26]

Radio-blackout storms are short-lived nuisances, though in today's world they can be expensive nuisances. Minor ones occur frequently—about 2,000 times per solar cycle, according to NOAA's figures. Extreme ones are rare, averaging less than one

---

[26] See http://www.swpc.noaa.gov/NOAAscales/.

per solar cycle. Both are produced by the impact of high-intensity electromagnetic radiation, mostly ultraviolet light, on the ionosphere—a layer of the upper atmosphere extending from about 50 to 1000 kilometers into space. It's an effect long known to AM radio stations and shortwave radio enthusiasts, who use the ionosphere to bounce signals beyond the curve of the horizon. Solar storms can change that layer's height and thickness, with the result that for minutes to hours; long-distance communications can be anything from erratic to bizarre to impossible.

But it's not just your local radio station whose signals can be affected. In major radio-blackout storms, satellite communications are also impeded, and in today's world, these are nearly ubiquitous. "Every time you go to a gas station and purchase a gallon of gas with your credit card, that's a satellite transaction," Bogdan says.

Global Positioning System (GPS) signals can also be affected—which isn't surprising because GPS is dependent on radio transmissions from satellites far above the ionosphere. These signals are basically extremely accurate timing signals, which means that anything that alters their time of flight (such as a change in the ionosphere) can interfere. In fact, Baker says, the Halloween Storms of 2003 did just that, making GPS unavailable for aircraft navigation for 30 hours.

This interference with GPS was a big-enough nuisance in 2003 when, Baker notes, it was still a relatively young industry, worth $13 billion annually. Now, it's rapidly approaching the trillion-dollar mark.

Most of us associate GPS with that sometimes-irritating voice on our smart phones or automobiles that tries to tell us the way to the mall. But GPS has a lot of other uses. For example, Bogdan says, financial transactions now use GPS signals as highly accurate clocks for time-stamping transactions. Knock out GPS for a day, and the banking system and stock markets go with it.

Julia Slingo, chief scientist at Britain's MET Office (the UK's national weather service), compares all of this to what happened to Europe when Iceland's Eyjafjallajökull volcano erupted in 2010, bringing European air traffic to a halt for the better part of a week. It's not that Icelandic volcanoes had never before erupted, she notes. It was just that it had never before happened during the modern aviation era. "We are increasingly vulnerable to things that are really quite run-of-the-mill," she said in an invited lecture at a meeting of the American Geophysical Union, shortly after the eruption. "It's a natural hazard that only had an impact because of our changing vulnerability. We'd been complacent, without realizing that on our doorstep was something that could cause us problems."

The same, she said, applies to our vulnerability to solar storms—a vulnerability that has done nothing but increase since the 2003 Halloween storms. "Our dependence on global communications has grown, as we've been sitting here complacently," she said. "We are increasingly vulnerable in this interdependent environment in which we live."

Nightfall

Interference with communications is by no means the biggest risk. Radio blackouts are temporary. The moment the storm abates, they fade and pass. Radiation is more dangerous.

People on the ground are safe, but radiation storms pose

serious risks to astronauts caught outside of their radiation shelters. In the biggest radiation storms, NOAA even warns of risks to passengers on high-flying airplanes, especially on flights going over the poles, where the Earth's magnetic field tends to channel higher levels of fast-moving particles. In addition, poorly shielded satellites can be at risk for equipment damage, ranging from compromised solar panels to computer-chip problems causing an inability to orient properly, possibly putting them out of contact with the ground forever.

But the scariest of all storms are geomagnetic ones, the massive disturbances that can send currents zinging through telegraph wires, pipelines, or electrical wires. The big fear here is what might happen to the electrical grid if power surges blow out giant transformers. In the U.S., Baker says, the grid is so thoroughly interconnected that a cascading power surge could take out hundreds of them, shutting down power to millions of people. Especially vulnerable, he says, are the 130 million residents of the eastern half of the country, where the grid is particularly interconnected.

Unlike the power relays taken out in the 1989 Quebec Blackout Storm, blown-out transformers can't be brought back online in a matter of hours. They are big, costly, and not easily replaced, especially if hundreds have been destroyed. "They don't have a lot of these on the shelf," Cliver says. "They'd have to gear up a Manhattan Project to make them quickly." Meanwhile, 130 million people in America, plus millions more in Europe and Asia, would be without power.

It sounds like a great scenario for a science-fictional disaster story. . Except...it seems so prosaic. Spoiling food. People trapped on elevators. No lights. No computers. No jobs. A

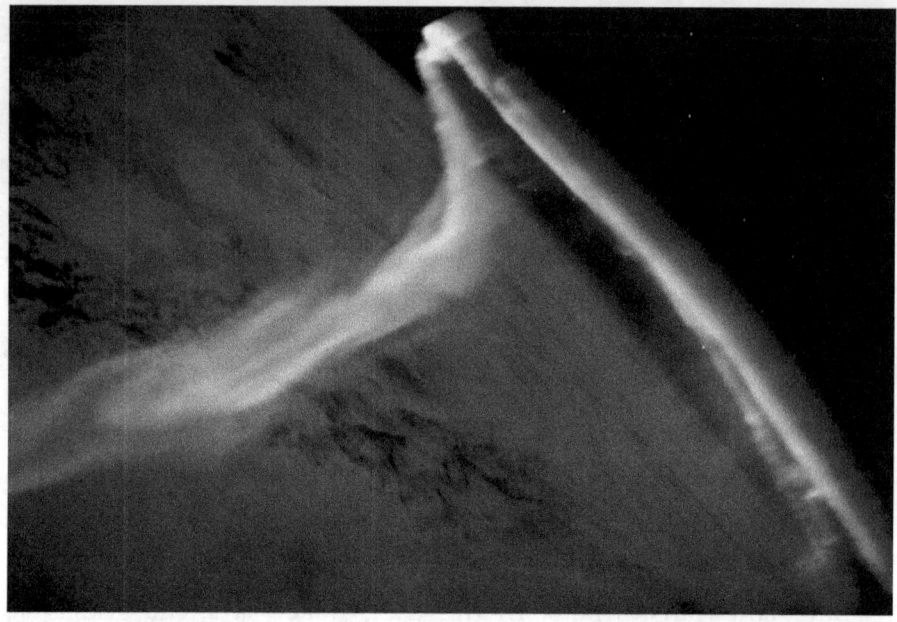

Southern lights from space. Credit ISS crew, mission 23.

sudden, enormous enormous demand for gasoline-powered generators and for axes to split wood, a desperate hunt for seeds for gardens. But as it drags on and on and authorities struggle to supply water, food, sewage treatment, medical services, and other basic needs, the situation could become increasingly bleak. When the National Research Council report first came out, *New Scientist* magazine envisioned the aftermath this way: "A year later and millions of Americans are dead and the nation's infrastructure lies in tatters. The World Bank declares America a developing nation. Europe, Scandinavia, China and Japan are also struggling to recover..."[27] Or, as John Kappenman, a power industry analyst at Metatech Corporation in Goleta, California,

[27] Michael Brooks, "Space Storm Alert: 90 Seconds from Catastrophe," *New Scientist* 23 March 2009.

told the writer of that article: "It's just the opposite of how we usually think of natural disasters. Usually the less developed regions of the world are most vulnerable, not the highly sophisticated technological regions."

"Off-Switch" Brinksmanship

The long-term solution, obviously, is to rebuild the power grid to reduce its vulnerability to cascading current surges that could bring everything down in a single blow. That's what Sweden has been doing, says Helena Lindberg, director general of the Swedish Civil Contingencies Agency, who adds that the power outage during the 2003 Halloween Storms actually represented the system functioning as intended. "We have invested quite a lot in protecting our grid," she said at a 2011 meeting of the American Association for the Advancement of Science.

But even without such changes, there are things we can do. Geomagnetic storms don't arrive completely without warning. Not only are big flares visible hours before any associated CME might hit us (although the link between the size of the flare and the size of the CME is not well understood), but a satellite called the Advance Composition Explorer (ACE) currently sits 1.5 million kilometers from the Earth, on a direct line between us and the Sun. "Any storm moving toward us from the Sun has to pass across that satellite," Bogdan says. That doesn't give us a lot of warning, but even for a 14½-hour event like the 1972 CME, it would be nearly ten minutes.

What would we do if we had ten minutes' warning of a Carrington-level CME heading our direction?

Panic, might be one reaction. But there are ways to minimize the damage. One of the big ones would be for power companies to protect transformers by taking them offline before the storm strikes. That would produce blackouts, but these would be of short duration compared to what might otherwise happen. "The good news is that these storms tend to pass after a couple of hours," Bogdan says.

There are just two hitches. One is that the ACE is the only satellite of its kind, and it's approaching 20 years old. "When people ask me what keeps me awake at night," Bogdan said in 2011, "it's whether or not that satellite will be running the next morning." (A backup satellite, the Deep Space Climate Observatory, or DSCOVR, is scheduled for launch by the time this book prints. It was built years ago, but appears to have been mired in climate-change politics due to the fact that the original mission was to look back at the Earth in order to make measures related to global warming. For updates, see www.nesdis.noaa.gov/DSCOVR/.)

Another problem is who makes the decision to shut down in marginal circumstances, where it might not be necessary. "If I shut down first, my rival makes money," says Stephan Lechner of the European Commission's Joint Research Centre Institute for Protection and Security of the Citizen. The result could be a game of economic brinksmanship, with the fate of the global economy at stake.

All of this gets easier if better forecasting gives us more time to decide what to do. On February 11, 2010, NASA launched the Solar Dynamics Observatory, which began observations in early 2011. The goal was to learn more about solar processes as the Sun moves deeper into the current cycle, with the hope of learning how to predict when and where solar storms might appear. But don't expect instant results. "We're back where weather forecasters were 50 years ago," Rodney Viereck, a physicist at NOAA's Space Weather Prediction Center in Boulder, Colorado, said at a geophysics meeting in 2010.

Nor should we take too much comfort from the fact that the current solar cycle appears to have been the least active in nearly 200 years. The cycle that produced the Carrington Event was also relatively inactive. "So it's not the case that times of modest sunspot number necessarily correlate with weaker geomagnetic storms," Baker says.

The Carrington event appears to have happened due to a rare concatenation of events that managed to focus unusually intense solar energy on the Earth—a combination that added up to the ultimate solar storm. Rare is good. But could it happen again?

"Yes," Tsurutani said without hesitation, all the way back in 2004.

How soon?

Nobody knows.

GEOLOGY, GEOHISTORY & PSYCHOHISTORY

Nobody knows why J Harlan Bretz ("J" was his full first name) first came to the scablands of eastern Washington. It was 1922 and he and a group of University of Chicago students were heading for the glaciers of the North Cascades when for some reason they stopped. Whatever the impetus, it was a decision that would reshape geology and ultimately play a major role in our understanding of Mars. First, though, it would get Bretz snubbed by his profession in a debate with intriguing parallels to Isaac Asimov's classic *Foundation* series of science fiction stories, even though Asimov's stories weren't written until more than two decades later.

For both Bretz and Asimov, the issue was one of gradualism vs. rapid change. In *Foundation*, the gradualists were engaged in the hypothetical science of psychohistory, which held that while individual human actions are variable, large groups respond to social, psychological, and economic forces that make it possible to predict the future, at least in broad terms. But Asimov's story also had a character called the Mule, a mutant human with abilities that allowed him, unpredictably, to bend the course of history.

Intellectually, Asimov appears to have sided with the gradualists but was aware that unexpected events (such as the birth of the Mule) could undermine their predictions. Bretz's colleagues, (known in geology as uniformitarians), weren't as open-minded.

The scablands is a region where the rich soil of Washington's wheat country is interrupted by raw gashes carved into the dark, basaltic bedrock. Bretz would later describe them as "wounds only partially healed—great wounds in the epidermis of soil with which Nature protects the underlying rock." They baffled him. Summer after summer he returned, endlessly hiking: mapping, map-ping, mapping. The more he mapped, the odder the area looked. There were places where, in the words of Bretz's biographer, John Soennichsen, the hills seemed "V-ed off" at one end, as though they'd once been half-eroded islands.[28] Elsewhere, valleys seemed half-formed, as though erosion had started, then abruptly ceased.

There were also giant basins that looked, for all the world, like river-bottom potholes. "But potholes are typically a few meters across at most," says Victor Baker of the University of Arizona. "These are a hundred meters across and maybe fifty

---

[28] *Bretz's Flood* (Sasquatch Books, 2009).

meters deep." It looked like the aftermath of an immense flood. A Noah-scale flood. The type of thing Bretz's fellow geologists would instantly pooh-pooh.

Uniformitarianism says that geological processes occur slowly, over the course of what author John McPhee once dubbed as "deep time"—in manners we see happening around us, day-to-day. The Grand Canyon, for ex-ample, wasn't dug by one catastrophic gully-washing rain: it was formed milli-meter by millimeter, via millions of years of perfectly ordinary spring floods. To the geologists of Bretz's

Photo by Dr. Julian Goldsmith

era this principle had the force of dogma. In part, that was an outgrowth of science-and-religion debates in which scientists didn't want to resort to miraculous-sounding cataclysms to explain what they saw. But it was more than that. Many early geologists were devout Christians. "I think that distinction has been very much overplayed," Baker says. Rather, he thinks uniformitarianism was a reaction to the type of scientific snobbery that casts field sciences like geology as poor relations to laboratory sciences like physics and chemistry. To counter this, geologists wanted a fundamental principle on which to base their work. "Uniformitarianism was to make the science respectable compared to physics," Baker says. "People thought you needed a principle to ground the science in something strong."

Soennichsen calls it an inferiority complex. "[They] weren't guys in lab coats with test tubes—they were chipping away at rocks, which a five-year-old can do. The inferiority complex led them to latch onto this fundamental principle."

As I was writing this article, I took a hike in Oregon's Columbia River Gorge. A 4,000-foot-deep gash through the Cascade Mountains, it's a place all of Bretz's water had to flow through en route to the sea. Now that I knew what to look for, there was plenty of evidence there, as well. On my hike, I found water-rounded pebbles hundreds of feet above the present-day river, with no creek in the vicinity to explain them. It's also a realm with more than 75 named waterfalls, some hundreds of feet high, created when enormously high water scoured out the Columbia River channel, leaving its side-creeks in perched valleys, high above.

Back in Portland, my own neighborhood also fits Bretz's theory. Not far away is an extinct volcano called Rocky Butte. Its east side—upstream at the time of Bretz's flood—is steep, eroded, stripped of soil. The west side has more soil, and downstream extends an area known as Alameda Ridge. In aerial views, the ridge has the classic shape of gravel bars I've encountered on many a canoe trip. But it's a gravel bar writ large, 200 feet above the present-day river.

One of my best friends once lived near the ridgecrest where, digging a garden one spring, we kept coming up with rocks the size of hubcaps. Not only were they obviously water-rounded, they were made of granite or granite-derived metamorphics. But there's no granite bedrock in our part of Oregon: her garden was full of chunks from somewhere far away, washed downstream by Bretz's flood and deposited in a gravel bar 200 feet high. I have a book on my shelves about Bretz and his "humongous" flood.[29] *Humongous* barely begins to describe it.

Even Bretz was a reluctant catastrophist. "He kept trying to explain that he'd gone back and looked again and again, and the only way [he could] account for this is [by] huge quantities of water," says Soennichsen. But he also had a problem: he couldn't figure out where all of that water came from.

If you're a geologist looking for a mechanism that might create big floods, one place to look is Iceland (see chapter 15). The Icelandic floods are called jökulhlaups, and they're common enough to satisfy even the most ardent uniformitarian. The most recent, in 1996, produced three cubic kilometers of water in two days, crushing that country's largest bridge like used tissue paper.

Jökulhlaups occur when volcanoes erupt beneath glaciers. Meltwater collects in ponds beneath the ice until the ice lifts, giving the water a way out. Then, poof, no more bridges, farms, or anything else that happens to be in the way. But no jökulhlaup has ever come close to the scale needed to account for Bretz's scour channels. Nor could he find a place where big volcanoes might have erupted beneath ice—and fully aware of jökulhlaups, Bretz looked for signs that this is what had occurred.

Eventually, in 1927, he was invited to Washington, D.C. to

---

[29] David D. Alt, *Glacial Lake Missoula and Its Humongous Floods* (Mountain Press Publishing Company, 2001).

speak before an elite gathering of geologists. At last, he must have thought, he'd been given a chance to prove his case. "He did a great deal of preparation," says Soennichsen. "He had all sorts of visuals, maps. It would have taken an hour-and-a-half or more to make the presentation. "But he'd walked into an ambush. The assembled scientists had no intention of being persuaded and one after another they rose to squash his outrageous theory. Caught by surprise, Bretz made a poor defense, then, depressed, boarded a train back to Chicago. He would never again return to the scablands.

The problem with grand theories like uniformitarianism is that they blind you to what the data itself might be trying to say. "The grounding for a science is not a principle," Baker says. "The grounding is that you are open to what nature has to tell you. If you dismiss something as impossible, you will not learn anything about it. It's like being a detective. If you ignore a clue, it's the critical one." In physics, scientists generally form hypotheses, then conduct experiments to test them. But geology is an observational science, in which experiments involve forming a hypothesis, then going out into the field to see whether you can find other things that are consistent (or inconsistent) with it. "Again, it's like a detective at a crime scene," Baker says. "When they're on the right track, things begin to fit a pattern that wasn't obvious before."

It appeared that the uniformitarians had won. But Bretz had a secret supporter, a man named Joseph T. Pardee. During Bretz's Washington D.C. presentation, Pardee reputedly turned to someone sitting near him and confided: "I know where

Bretz's water came from."

Bretz had figured that the water had to have had something to do with melting Ice Age glaciers, not far to the north, but Pardee believed the source was farther east, in Montana. The uniformitarian/catastrophist divide was so strong, however, that he wasn't willing to risk his career supporting a catastrophist. Not yet. Thirteen years later, however—in 1940, just as he was retiring—Pardee dropped his bombshell in a vaguely titled paper about ripple marks in the bed of glacial Lake Missoula.

It wasn't the type of title designed to draw publicity. Lake Missoula was an Ice Age lake about the size of one of the Great Lakes that had formed in western Montana when ice blocked what is now the Clark Fork River, causing water to impound up to 2,000 feet deep. Geologists had long known about it: Pardee himself had mapped its shorelines in 1909 and published his findings in 1910. But he'd not mentioned the ripple marks. Those he'd kept to himself.

They were a lot like the ripples ordinary streams leave in sandbars, but they were enormous: 30 feet tall, a mile long, and spaced at intervals of 200 to 300 feet. Such features could only be created by an enormous current: the type of current that could only be produced if the entire lake emptied, practically overnight.

Today, scientists know that ice dams are notoriously unstable. When the one that formed this lake broke, they estimate, water roared down the canyons at 50 to 60 miles per hour. It was, quite simply, one of the greatest flash floods of all time. Pardee never bothered to say where all of that water went. He didn't need to. Anyone who'd heard of Bretz knew.

Today, scientists have found signs that ancient Lake Missoula may have formed and drained dozens of times over the course of several thousand years. Others have found traces of additional super-floods in Canada, Siberia, Mongolia, and Europe. Then in 1973, Mariner 9 returned the first good, close-up photos of Mars. One of the people who saw them was Baker, who'd done his doctoral research studying the Eastern Washington scablands. What he saw looked familiar. "It was pretty obvious that the features were similar," Baker says. Today, most planetary scientists believe that Mars, too, had enormous floods sometime in its history, though they are still searching for the source of the water.

Bretz himself lived to 98, old enough to watch the Martian discoveries and be pleased at hearing the newly discovered features referred to as scablands. Better yet, in 1979, the Geological Society of America (the very organization that had arranged the D.C. meeting at which he had been so thoroughly excoriated) awarded him its highest scientific honor, the Penrose Medal. Bretz reportedly had one complaint: "All my enemies are dead, so I have no one to gloat over," he said.

Today's question isn't whether there was a big flood in the scablands: it's whether even Lake Missoula was big enough to account for everything Bretz saw. Today's "outrageous" hypothesis, Baker says, is that additional water came from beneath the ice caps—like a jökulhlaup but without a volcano.

The idea stems from findings in Antarctica, where scientists have discovered large subglacial lakes, one of them enormous. More recently, they've discovered that water can flow from one lake to another via subglacial channels. Could a similar lake, perhaps filling part of the great trough of British Columbia's

Okanogan Valley, have belched water, big time?

The idea, Baker admits, is a long shot. "A lot of people would have thought the ice sheet was stable," he says. But, he adds, "it's a principle of science that if you dismiss something as impossible, you will not learn anything about it."

That sounds like a pretty good principle for any scientific endeavor. And in the wake of Bretz's vindication, you might think the old gradualism/catastrophism debate would be finally put to rest. But that's not the case. It's alive and kicking in history and archaeology. There are plenty of examples, but the one we'll look at for the rest of this chapter involves the eastern Mediterranean.

Big Bang in the Aegean

For hundreds of years, the Minoan culture thrived on the island of Crete, dominating the entire region. Its most famous ruin is the palace of Knossos, whose warren of passageways probably gave rise to the legends of the Labyrinth and the Minotaur. But then, about 3,500 years ago, Minoan civilization came crashing down. The ruins show clear evidence of fire and violent destruction, but what was the cause? Some early excavators suggested a cataclysmic event, but the idea fell into disrepute. And by the late Twentieth Century, says Amos Nur, a geophysicist at Stanford University, archaeologists tended to reflexively blame such cultural collapses on internal decay that might also have opened the door to invasion.

Nur attributes this attitude to one of the Twentieth Century's leading historians, Arnold J. Toynbee, who in 1939 argued that for the twenty-plus civilizations he'd examined, the cause of collapse was internal, not external. "The breakdowns of civilization are not catastrophes of the same order as famines and floods and fires and shipwrecks and railway accidents; and they are not the equivalent, in the experiences of bodies social, of mortal injuries inflicted in homicidal assaults," Toynbee wrote.[30] In other words, his view was that you can't kill a culture with a single blow (unless, perhaps, it is already dying of internal causes). It's basically, a gradualist view, not all that different from that of the uniformitarians of Bretz's time.

Just as Bretz ran into trouble for proposing his gigantic flood, historians and archaeologists who suggest that civilizations may have fallen due to natural cataclysms have been mocked and may well have damaged their careers. The fear is so deep-seated, Nur says, that when one archaeologist suggested that an earthquake played a role in one city's collapse, she felt compelled to add: "Archaeologists of my generation…were brought up to view earthquakes, like religion, as an explanation…to be avoided if at all possible."[31]

But the Aegean and eastern Mediterranean lie in an extremely active tectonic zone near the boundaries between the African Plate, the Eurasian Plate, the Arabian Plate, and the Anatolian (Turkish) Platelet. As much as five percent of all the earthquake energy released on Earth is concentrated in this zone, says Jelle Zeilinga de Boer of Wesleyan University in Middletown,

---

[30] A. J. Toynbee, *Study of History*, Vol. 4, as quoted by Nur in *Apocalypse: Earthquakes, Archaeology, and the Wrath of God,* (Princeton University Press, 2008).

[31] Elizabeth French, "Evidence for an Earthquake in Mycenae," in *Archaeoseismology*, 51-54, S. Stiros and R. E. Jones, Eds. (1996), as quoted by Nur.

Connecticut.[32] There are also volcanoes, the most famous of which is (or was) Thera, in the Aegean archipelago of Santorini.

Thera lay about 70 miles north of Crete. Sometime about 3,600 years ago, it erupted in the biggest volcanic blast in the history of civilization: an order of magnitude bigger than the one that in 1883 destroyed the Indonesian island of Krakatoa. It's a little hard to determine when, precisely, the eruption occurred, but Zeilinga de Boer notes that sulfur deposits in Greenland ice cores date the event to about 1645 B.C.E. Tree-ring data from California, Turkey, Sweden, and Ireland, he adds, all show a decade-long period of severe global cooling at about the same time, while Chinese records show "yellow fogs," probably from sulfuric acid, as well as frost in July.

Nobody doubts that Thera produced a big, nasty bang. But the demise of Minoan civilization is normally dated at about 1450 B.C.E., nearly two centuries later. For gradualists, that's proof that in the long haul, volcanoes are irrelevant. Healthy civilizations weather whatever disaster befalls them; unhealthy ones are collapsing anyway.

Zeilinga de Boer has a different theory. He calls it the "vibrating string" and argues that natural cataclysms can have impacts that ripple through history for years, decades, and even centuries. Crop failures can lead to famine and disease; ecosystem changes can alter economies and ultimately, cultures. It's a bit like chaos theory's famous "butterfly effect" or Ray Bradbury's classic short story, "A Sound of Thunder." Chaos theory holds that incredibly minor changes in "initial conditions" can have enormous long-run impacts: analogized to the

---

[32] Jelle Zeilinga de Boer and Donald Theodore Sanders, *Volcanoes in Human History* (Princeton University Press, 2002).

hypothetical butterfly flapping its wings in China that alters North America's weather, weeks later. Bradbury's story, (which also involved a butterfly, this one accidentally stepped on by a time traveler), even suggested that over the course of time, the death of even a single insect could have enormous impacts.

Returning from fiction to real history, Zeilinga de Boer points to a temblor that in 464 B.C.E. damaged the ancient Greek city-state of Sparta. The Spartans were a relatively small warrior caste supported by a much larger slave population. The earthquake, whose magnitude is estimated to have been 7.2, had an epicenter directly beneath the city. It knocked down many buildings and killed many soldiers, as well as many women and children, depleting the ranks of the next generation's soldiers. To compensate, the Spartans had to fill out their army with slaves, "which wasn't so good," Zeilinga de Boer says, "because those people weren't as interested in fighting for Sparta."

Sparta limped on for several decades but eventually was defeated: one of the most momentous events in ancient history because it opened the door for the rise of Athens and the ensuing Golden Age of Greek culture. Without the earthquake, Zeilinga de Boer argues, the history of western civilization might have been unimaginably different.

Nor are such changes limited to ancient history. In 1972, Zeilinga de Boer says, a magnitude 6.2 temblor in Managua, Nicaragua precipitated revolution by exposing the corruption of the government, whose officials used relief money to line their own pockets. That led to the leftist Sandinista regime, to which the United States reacted, with long-term effects on U.S./Central American policies. "Disasters have long-ranging after-effects,"

Zeilinga de Boer says. "If a major earthquake occurred in California, there would be so many social and economic impacts that people couldn't comprehend it."

So, what did happen in ancient Crete? One possibility is that we've got the date for the collapse of Minoan civilization wrong and it actually occurred earlier, in the immediate aftermath of Thera. But there's no obvious reason to suppose that the accepted date (obtained by matching up pottery styles in Crete and Egypt) is wrong. Sure, dating such things isn't rocket science, but two centuries is an enormous error. More likely, if Thera did deliver the mortal blow, it took a long time to play out.

Zeilinga de Boer's hypothesis is that the eruption and associated earthquakes damaged Minoan cities, while the tsunami ruined harbors and boats, vital to a seafaring civilization. This opened the door for Mycenaeans (Greeks) to move in, as the surviving Minoans dispersed. In an interesting side effect, he argues, the Minoan written language was adopted by the Greeks. Pre-Thera, they had no writing. Post-Thera, they did. That ripple, far down Zeilinga de Boer's vibrating string from its initial source, might mean that the very philosophy and science of Western Civilization owes its genesis to a single volcanic blast.

Mycenae and other early Greek cities were not themselves immune to destruction. In 1993, Robert Drews listed 47 major archaeological sites in Greece, Asia Minor, Syria, and Israel that showed clear signs of collapse, fire, and/or abandonment somewhere between 1225 and 1175 B.C.E. "Within a period of forty of fifty years...almost every significant city or palace in the eastern Mediterranean world was destroyed, many of them never to be occupied again," he wrote.[33] One, ironically, was Mycenae itself.

Traditionally, blame for all of this destruction has been laid at the feet of mysterious "Sea Peoples" who attacked, pillaged, then vanished. There's just one problem: nobody has a clue who they were or why they didn't hang around to enjoy their spoils.

In 1993 Nur visited the ruins of ancient Mycenae, where he was immediately impressed by the fact that the city's fortifications had been built atop an obvious fault scarp. Obvious to him, at least. To the ancient Greeks, it would have looked like any other cliff, which they happily incorporated into their fortifications. Nor, he realized, had archaeologists appreciated the significance of that cliff, whose surface was smooth, polished: a clear sign of geologically recent seismic activity.

Thanks to their anti-catastrophist mindset, Nur adds, archaeologists haven't been all that good at actually looking for evidence of earthquake damage: unfortunate, because earthquakes and marauders leave different types of ruins. If a fault line cuts directly across a wall or fence, for example, movement of the fault will put a distinctive kink into it. And that's not the only smoking gun for earthquake damage. If raiders destroy a temple by pulling down its columns, the columns will fall every which-way, depending on the angle at which they are pulled. But when columns or pillars are knocked

---

[33] R. Drews, *The End of the Bronze Age: Changes in Warfare and the Catastrophe ca. 1200 BC.* (1993), as quoted by Nur, *supra*. As the title indicates, Drews believed that improved military techniques were the primary reason this occurred.

down by earthquakes, they tend to fall in parallel rows. In addition, crushed skeletons indicate that something brought everything crashing down a good deal more quickly than warfare was capable of in the millennia before gunpowder. Nur believes that Drews' cities were actually devastated by a "storm" of earthquakes that struck one after another (though war may also have played a role, as rival kingdoms took advantage of the holes knocked in their enemies' defenses).

One of the arguments that long led archaeologists to dismiss earthquakes as a crackpot explanation is that Drews' devastated cities are spread over a 900-mile radius, a vastly larger area than could be affected by a single earthquake. But we now know that earthquakes can occur in sequence, as stresses shift along fault lines. That's exactly what happened in Indonesia in the aftermath of the mammoth Sumatran earthquake of December 26, 2004: barely three months later, an adjacent segment of the same fault was hit by another gigantic earthquake. And in Turkey, from 1939 to 1967, a string of seven earthquakes (magnitude 5.6 and larger) progressed westward along 500 miles of the North Anatolian Fault, south of the Black Sea—just one of several faults that might have contributed to Nur's 3,200-year-old earthquake storm.

Ultimately, all of this brings us back to Asimov, chaos theory, Bradbury, and *Foundation*. From a theoretical perspective, many of us, like Toynbee, Bretz's critics, and Asimov's psychohistorians, are inherently drawn toward gradualism. After all, it offers the promise that if we can but tease out the rules by which the world works, then past, present, and future events are well-ordered: predictable and under control. But as readers, we are biased toward Bretz's flood, the Mule, Nur's earthquake storm. Predictability makes for poor stories. In the real world, though, it may be that Bradbury and chaos theory have it best. The laws governing the world may be well ordered, but maybe they are so incredibly sensitive to small effects that predictability is, to all intents and purposes, impossible.

Scientifically, I like grand unifying theories like gradualism. In fiction? Give me a little bit of vibrating-string chaos, any day. To a large extent, in fact, the entire field of alternate history is based on its own version of Zeilinga de Boer's vibrating string concept. Because if he's right, the story options are virtually unlimited.

FROM ATLANTIS TO CANOE-EATING TREES

Geomythology Comes of Age

Long ago, according to Indian legends of the inland Pacific Northwest, the Great Spirit's twin sons, Wyeast and Pahto, dwelt on opposite sides of the big river the whites would one day call the Columbia. For many years they lived in peace. But then, as is the way with young men, they quarreled for the attention of a beautiful woman known as Tah-one-lat-clah. Soon, Wyeast and Pahto were hurling rocks and fire at each other, scorching the land and frightening its residents.

Learning of the commotion, the Great Spirit came back to restore order, forcing the brothers to apologize and promise to quit fighting. As a sign of the truce between them, he built a beautiful stone bridge across the river, not far from the place where the whites would later build Bonneville Dam. Then he went away again. For a while peace reigned, but the brothers soon forgot their promise and resumed their quarrel. Tah-one-lat-clah tried to intervene but was severely burned in the fray. The bridge was destroyed, and the brothers, dismayed at what they had done, withdrew to the locations where they reside today as the mountains later to be called Adams and Hood. Tah-one-lat-clah, now known as Mt. St. Helens, also moved away, seeking a place to hide, far from other mountains. There she nursed her wounds, and there she remained, even after the Great Spirit returned to heal her disfigurement and restore her to her former beauty.

It's a great story, though this version has probably been somewhat Paul-Bunyanized by white missionaries who collected it in the Nineteenth Century. But it's just a myth, right? A tale to be told to children and maybe teach a lesson about sibling rivalry?

Maybe not, say geologists. Such myths were once discounted, but these days, scientists are paying more attention. There's even a new field called geomythology that draws on everything from Aztec legend to Bible lore in an effort to better understand the Earth's turbulent past by correlating old stories to geological events.

Let's take another look at the Bridge of the Gods legend. As far back as 1805, Lewis and Clark knew there was something odd about that part of the Columbia. Approaching from upstream, they found the river curiously sluggish, with deep, calm waters in which the boles of dead firs rose from the bottom like drowned sentinels, 20 feet tall. Aware the water couldn't have been that high when the trees were alive, Clark figured something must have dammed the river. When the expedition reached the giant boulder field that (unknown to them) the Indians interpreted as the ruins of the Bridge of the Gods, Clark speculated it must have been formed by a gargantuan landslide, recently enough that the submerged tree trunks hadn't had time to rot.

As it turns out, he had everything right but the date. "They thought it occurred 20 years [before], but it was hundreds," says John Jengo, a geologist who's studied the Lewis and Clark expedition's records. "But there was no way for them to know that."

The Indian legend actually matches the region's geological history quite well. Mt. Adams has frequently vented steam and Mt. Hood had a series of eruptions several hundred years ago. Mt. St. Helens also went through a major eruptive phase from 530 to 350 years ago, one that began with the mountain blowing its top in the 1480s (determined by growth-ring dating of trees buried in the ash). Ensuing eruptions then slowly rebuilt it from an ugly, stumpy pyramid to the elegant cone that persisted until its famous 1980 explosion. "[These changes] would have been very noticeable," says Peter Frenzen, monument scientist at Mt. St. Helens National Volcanic Monument.

According to Pat Pringle, an earth science professor at Washington's Centralia College, the latest radiocarbon studies on the submerged tree trunks indicate that the Bridge of the Gods landslide occurred somewhere between 1435 and 1455. That's several decades before Mt. St. Helens' 1480s blast, but the events occur in the right order and in close-enough sequence for storytellers to have combined them into a single tale, especially since the story doesn't indicate how many years elapsed between each of its events.

Troy and the "Telephone" Game

Geomythology may be a new science, but it has precedents in archaeology, which has been discovering truths in ancient stories since at least the 1870s, when Heinrich Schliemann found the ruins of Troy, the city besieged and defeated in Homer's *Iliad*. Archaeologists have also found many biblical sites once believed to be nothing but myth. Today, biblical archaeology is a flourishing field, complete with its own journals.

But geologists have been slower to take the hint. Partly that's because many of the old stories sound so fantastic. Also, our culture is steeped in the concept of the "telephone game," in which a phrase is whispered to one person, who whispers it to another, and then another...until the statement reaching the end of the chain is barely recognizable. "Eliza has the flu," for example, could easily emerge as "Arise and get glue." Or worse. It makes us suspicious of any information passed along in this manner.

But we are not an oral-tradition-based culture. We're poor storytellers and worse listeners, with a bad tendency to assume everyone else has always been just like us. From everything anthropologists have been able to determine, the ancients were very good at passing on stories by word of mouth. And at least as importantly, they were a lot better than we are at paying attention to the world around them. "The ancients, because they were closer to nature, were very observant," says Jelle Zeilinga de Boer of Wesleyan University in Middletown, Connecticut.

Just how observant came to the fore in 2008, regarding a passage in *The Odyssey*. For those who've forgotten their Homer, *The Odyssey* is the story of Odysseus's multi-year effort to get

114

home to Greece after the fall of Troy. Eventually he makes it there only to discover that his wife, Penelope, is besieged by a host of suitors who presume him to be long dead and want not only her but his lands. Very much alive and even more enraged, he slaughters them all, but not before Homer describes the Sun as having "perished out of heaven" and an "evil mist" spreading across the world.

This passage is clearly a prophesy of doom. But in a paper in the June 24, 2008 issue of the *Proceedings of the National Academy of Sciences*, Marcelo Magnasco, a physicist at Rockefeller University, teamed up with Constantino Baikouzis of the Astronomical Observatory in La Plata, Argentina to argue that it also describes a total eclipse of the Sun—one that actually happened in about the right era of Greek history.[34]

Magnasco and Baikouzis weren't the first to hit upon the eclipse theory. More than 80 years ago, other astronomers went so far as to calculate that if this was indeed what Homer was describing, it had to have been on April 16, 1178 B.C.E. because that was the only total eclipse visible from Greece during the Homeric era. It even occurred at noon, as described in the poem. But Magnasco and Baikouzis went further and looked for other astronomical references in *The Odyssey*, finding several. Some related to the phase of the Moon (which must be new at the time of a solar eclipse). Others described Odysseus steering his boat by the positions of the constellations Bootes and the Pleiades, 29 days before the slaughter. Both constellations are visible at sunset, something that only happens at certain times of year. There is also a reference to Venus as a morning star and a

---

[34] Baikouzis & Magnasco, *Proceedings of the National Academy of Sciences*, 105 (26) 8823-8828 (2008).

somewhat cryptic reference to Hermes (equivalent to the Roman god Mercury) taking a back-and-forth trip to the far west and turning back, 34 days before—something that Magnasco and Baikouzis hypothesized might be a description of the planet Mercury at the far end of one of its retrograde cycles.

Armed with these references, the two scientists then used planetarium-style computer software to simulate the Greek skies, night by night for 135 years, centering on the known date of the eclipse. What they found was that the descriptions of the Moon, stars, and planets only fit the story line on one occasion: a sequence that had the suitors being killed on—you guessed it—April 16, 1178 B.C.E.

Did Homer really include an actual eclipse in his story? Who knows. "The implication is that 'Homer' (in quotes as [he] may have been many poets) was aware of astronomical events occurring four centuries before the poem was cast in its current form," Magnasco told me, "and was interested enough in those events and knowledgeable enough about them to weave them into the narrative."

But that's Magnasco's cautious interpretation. It's also possible there is a great deal of history woven into both *The Odyssey* and its prequel, *The Iliad*. "It adds to the evidence that [Homer] knew what he was talking about," he told the Associated Press when his study came out. If nothing else, it proves Homer was well versed in astronomical phenomena far before his own time—in about the right era for the story he was telling. Fact? Fiction? Myth? A mix of all three? We may never know for sure, but Odysseus suddenly sounds like a much more real character than historians once believed.

RICHARD A. LOVETT
Crossing the Jordan

Let's take a look at another old story, this one from the Bible: the account of how the Israelites crossed the River Jordan into the Promised Land. According to the story, the Israelites, who'd been wandering the Sinai wilderness for two generations, finally moved into Canaan in harvest season, which in that part of the world came in late spring. That meant they hit the river at flood stage, an obvious problem. For three days, they camped on the bank. Then their leader, Joshua, announced it was time to cross, with priests leading the way, carrying the Ark of the Covenant. And the moment the priests' feet touched the river, the Bible says, the water stopped flowing:

> It piled up in a heap a great distance away, at a town called Adam in the vicinity of Zarethan, while the water flowing down to the Sea of the Arabah (the Salt Sea) was completely cut off. So the people crossed over opposite Jericho. The priests who carried the ark of the covenant of the LORD stood firm on dry ground in the middle of the Jordan, while all Israel passed by until the whole nation had completed the crossing on dry ground. (Josh. 3:16-17)[35]

The river stayed dry long enough for the people, including 40,000 soldiers, to cross over. Then, as soon as the priests stepped out, the water returned, flowing at flood stage, as before.

---

[35] Scripture taken from the HOLY BIBLE, NEW INTERNATIONAL VERSION®. Copyright © 1973, 1978, 1984 International Bible Society. Used by permission of Zondervan.

So what's going on? The two simplest answers are that God intervened miraculously or it's all just a myth. But it could be a little more complex. It's well known that the Jordan River valley is prone to landslides that can block the flow of the river. The most recent was triggered by an earthquake on July 11, 1927, which occurred near the site believed to be the one called "Adam," about 16 miles upstream of Jericho, and it blocked the river for 21 hours. Another five such landslides, all caused by earthquakes, are known to have occurred in the past 3,000 years, says Zeilinga de Boer.[36]

It's the type of thinking that fuels arguments between conservative Christians, atheists, and everyone in between— though even some Bible commenters take the earthquake theory seriously, noting that there's still plenty of room for a miracle if you want one.[37] The Israelites, after all, seem to have done an extremely good job of being in the right place at the right time.

Vanishing Cities

Another example is the tale of Helike, a coastal city in ancient Greece, which reputedly disappeared overnight after an earthquake in 373 B.C.E. According to the legend, the earthquake occurred on a winter night. By morning, everyone was dead and the city had sunk into the sea. Some people, in fact, think that this is the source of the legend of Atlantis, though as we'll discuss later, that legend might have a better

---

[36] Zeilinga de Boer & Donald Theodore Sanders, *Earthquakes in Human History: the Far-Reaching Effects of Seismic Disruptions,* (Priceton University Press 2005).

[37] An example is Roger Barrier, "Did the Israelites Really Walk Across the Water?" www.preachitteachit.org (2008).

source farther back in history.

Accounts of the demise of Helike are sprinkled through Greek and Roman writings. There are stories of a rescue attempt in which nothing was found but the tops of trees, rising above the water, and stories of travelers, years later, seeing the ruins beneath their boats. But could a city truly be swallowed by the sea? For years, Zeilinga de Boer says, geologists viewed that as "total nonsense."

But there are ways for cities to vanish. In 1692, a powerful earthquake shook Port Royal, Jamaica. The destruction wasn't total, but in places the vibrations turned sand to quicksand that swallowed buildings and people whole. (To see this phenomenon on a small scale, try tapping your foot gently on damp beach sand and watch the way it tries to flow.) And this process, called liquefaction, isn't the only way a town can disappear. In British Columbia, natives long told of a village called Kwalate, destroyed when a big chunk of a mountain fell into the sea, beside a narrow fjord. It sounded like another Atlantis myth, but the village was real and archaeologists have found it, determining that it was destroyed in the late 1500s. The culprit: a giant landslide from a 2,800-foot peak that dumped somewhere between 100 and 140 million cubic meters of rock into the water, according to Brian Bornhold of the University of Victoria. Minutes later the entire village, home to about 100 people, was gone, wiped out by a wave that might have topped 30 feet in height.

In 1988, a team led by Dora Katsonopoulou of the Helike Project[38] and Steven Soter of the American Museum of Natural

---

[38] www.helike.org.

History began a serious search for Helike, which they knew lay somewhere along the Gulf of Corinth, a narrow rift west of Athens that nearly slices Greece in half. Their story is an exciting exercise in false starts and dashed hopes.[39] They found ruins, only to determine that they were from a Roman city that postdated Helike. Then they found more ruins, but these were from a Bronze Age village dating back nearly 5,000 years.

Then, in 2001, Katsonopoulou and Soter finally found what they were looking for: a Greek city from the Fourth Century B.C.E. Never looted by treasure hunters, ancient or modern, it might be one of the greatest archaeological finds of recent times: a Pompeii, buried in mud.

In part, Katsonopoulou and Soter determined, Helike appears to have sunk due to liquefaction. But there also may have been a tsunami that ricocheted through the narrow gulf. Either way, an entire city had indeed vanished overnight. It may even have gone underwater as well as being buried in mud; coastal earthquakes can create lagoons a mile or two inland from the beach. The myth of Helike was no longer a myth.

But does that mean Helike was Atlantis? Probably not. The Atlantis myth dates from the writings of Plato in about 355 B.C., less than a generation after the destruction of Helike. It's hard to imagine that a real disaster like Helike's could have been so thoroughly mythologized so quickly. Besides, Plato said the story originated in Egypt then was told to a visiting Greek named Solon. Plato wrote that Solon was told that Atlantis had sunk 9,000 years before—roughly 11,000 B.C.E. by our calendar. There's little chance an advanced civilization could have existed

---

[39] For details, see Tom Gidwitz, "City of Poseidon," *Archaeology*, Jan/Feb 2004.

at that date, which is part of why many people have either discounted the story entirely or tried instead to link it to Helike, difficult as that might be.

Then, in 1969, Greek seismologist Angelos G. Galanopoulos suggested that Plato might have botched one little detail.[40] Maybe instead of being told about Atlantis when he was in Egypt, Solon *read* about it on an Egyptian scroll and mistranslated the date: not 9,000 years before, but 900, putting the date at about 1250 B.C.E. Plato also said Atlantis was 2,500 miles away from Egypt—thereby putting it somewhere in the Atlantic Ocean, beyond the Strait of Gibraltar. But suppose that was a similar misreading and should have been 250 miles? If so, we have a very, very good candidate for Atlantis, right in the eastern Mediterranean: the Aegean island of Santorini.

Santorini was part of a large volcano called Thera and was inhabited by outposts of the seafaring Minoan civilization, based in Crete (see Chapter 5). Thera has been active on and off for hundreds of thousands of years, most dramatically in a blast that was probably the largest volcanic eruption in human history. That blast, occurring in about 1600 B.C.E., blew a large part of the island into the stratosphere and may have set in motion a chain of events that took down the entire Minoan civilization.

The date of the eruption doesn't perfectly match Galanopoulos' 900 years but it's not wildly off, and the blast had to have had major effects throughout the eastern Mediterranean. "That eruption was a mega-event in the middle of a flourishing culture," says Floyd McCoy, a volcanologist at the University of Hawaii's Windward College. "I think it stuck as a myth, and the

---

[40] His book was *Atlantis: The Truth Behind the Legend* (1969).

myth that best seems to reflect that is Atlantis."

Sniffing Gas

Yet another Greek story is the tale of the oracle at Delphi. The oracle was a temple served by priestesses whose prophecies were held in high regard through centuries of Greek culture. According to Plutarch, the women would first descend to the temple basement and breathe sacred fumes emanating from a fissure in the temple floor. Only then would they make their pronouncements. But modern historians haven't believed it.

There's no doubt the temple existed, but historians have thought the entire thing, including the fumes, was an ancient hoax because Delphi didn't sit on a volcano that might have produced intoxicating fumes. Instead, the underlying rock was limestone.

Visiting the temple ruins in the 1980s and '90s, however, Zeilinga de Boer noticed two geological faults running through the area, intersecting directly beneath the ancient temple. Later, he and some colleagues took a better look at the rock and discovered that the limestone was rich in hydrocarbons. Suddenly, the ancient story made sense: perhaps the priestesses were indeed breathing fumes, but they weren't volcanic gases. Rather, they were hydrocarbons baked out of the underlying rocks by subterranean heat and finding their way to the surface through the faults.

Zeilinga de Boer and a team that eventually included an archaeologist, a geochemist, and a toxicologist, made tests on water drawn from nearby springs. In 2001, they announced their findings in the journal *Geology*.[41] The fumes at Delphi, they said, were a mix of several gases, but the most important was ethylene. In large doses, it has been used as a general anesthetic; in smaller ones, it can make you high. The priestesses might as well have been smoking peyote.

Today, there's not enough gas to notice except in lab tests. But that's not because it's not being produced in the underlying rocks, Zeilinga de Boer's team suggested. More likely, modern well drilling altered the water table, causing the gas to go elsewhere (and probably disperse, since today's residents don't appear to be getting high on it.)

Canoes in Trees

It's hard to say precisely when geomythology became a serious science. The link between Thera, Crete, and Atlantis was suggested as far back as 1939, but even as the theory spread, it was viewed as an exception: a few myths might be backed by something real, but that didn't mean the same applied to others. "Five years ago, I couldn't do this without destroying my career," the University of Hawaii's McCoy told me in 2006.

Perhaps the field has yet to truly make its mark as a science, but it hasn't been hurt by recent findings in the Pacific Northwest, where a few years ago researchers began to take seriously Indian legends of waves so big they changed the shapes of islands and killed people by throwing their canoes into trees.

Like so much of geomythology, it sounds rather fanciful— enough so that Brian Atwater, a U.S. Geological Survey geologist based at the University of Washington, says that the legends

---

[41] Jelle Zeilinga de Boer, et al, *Geology* 29.8, 707–711 (2001).

themselves weren't even the motivators for the studies that ultimately confirmed them. Rather, scientists were starting to become concerned about the possibility of mammoth earthquakes offshore from the Pacific Northwest.

What they knew was that there was a large subduction zone offshore, similar to the one that would later produce the tsunamis that devastated Indonesia and Japan. But the Cascadia Subduction Zone was thought to be inactive. Then, in the 1980s, geologists started finding signs of sudden land subsidence. Meanwhile, archaeologists were finding Indain fishing camps that had been overrun by waves. Growth-ring dating of trees killed by the event put it in 1699 or 1700—a date that corresponded to Japan's "Orphan Tsunami," which damaged coastal villages on January 27, 1700.

The Japanese had long known that tsunamis were triggered by earthquakes, but the Orphan Tsunami had struck with no accompanying temblor (hence its name). At the time, the Japanese were puzzled. Now we know it started in the Pacific Northwest then traveled all the way across the ocean to strike Japan. And if it could do that, it could certainly throw canoes into trees.

This discovery, Atwater says, provides an excellent opportunity for "calibrating" myth against reality. And what it shows is that, despite generations of retelling, the legends retained an amazing amount of truth. "Here's an event that occurred almost a hundred years before the first European contact, and about 150 years before ethnologists started to write down the traditions of these people," he says. "By then, smallpox and other diseases had wiped out large parts of the population that could be telling these stories, so what's left has to be only a fraction of what was around, but still, you have an event that's pretty well established from geology."

More recently, Lori Dengler, a geology professor at Humboldt State University, California, found something similar in the aftermath of the deadly tsunami that swept Indonesia in December, 2004. In most places, the wave caught people unprepared. But on Simuelue Island, so close to the epicenter that the wave struck within eight minutes, there were only seven confirmed deaths out of 78,000 people. The reason, Dengler says, is that the island had been hit by a similar event in 1907, with massive casualties. The survivors told their children, who told their children, etc., until, 97 years later, everyone still knew that when the earth shook, it was time to run for high ground. When asked where to go for safety, they all pointed to a hillside, about 100 feet above sea level. "That's about where I would have told them to go," Dengler says.

So, what's the future? Obviously, there are plenty more myths to be explored. Aztec carvings, for example, indicate a belief that the world was destroyed on four occasions: once by jaguars, once by hurricanes, once by fire, and once by flood.[42] Other than the jaguars, Zeilinga de Boer says, the Aztec's disasters are geological events that may have left identifiable traces. "I suspect that once geologists start getting core samples, the pieces will fall together," he says.

Other scientists are taking a literal view of the even more ancient myth of a massive flood, recounted in the Bible as the story of Noah. In 1997, marine scientists Walter Pitman and

---

[42] These carvings were the source of 2012's "Mayan Apocalypse" frenzy.

William Ryan of Columbia University shocked geologists by arguing that the melting of Ice Age glaciers, 7,500 years ago, caused water from the Mediterranean Sea to cut through what is now the Bosporus Strait into the Black Sea, raising water levels by several hundred feet and inundating vast tracts of land.[43] Describing these arguments in detail is beyond the scope of this chapter, but they are the stuff of continued debate among geophysicists.

In general, this appears to be how the field of geomythology is developing. Few myths have led to new geological discoveries; rather, the geology has helped firm up the history surrounding them. "We find things and then go back to the mythology and say, 'My gosh,'" says McCoy. The next step, he believes, is to start parsing old legends for clues to previously unrecognized geological events. "I think that's just around the corner," he says.

---

[43] W.B.F. Ryan, *et al. Marine Geology*, 138:119–126 (1997).

DIVIDING THE INDIVISIBLE
The History of *Atomos*

Consider the most solid object you can imagine. Granite? The steel arch of the Golden Gate Bridge? Then delve deeper into it. What you'll find, we all know, is space. Lots and lots of empty space. Electrons, protons, neutrons, and weirder things with names like charm quarks or the as-yet hypothetical preon, spinning and bumping in a whorl of mostly nothing. Crunch not just the Golden Gate Bridge but the entire city of San Francisco down until all the bits touch each other, and all you'd have is an infinitesimal speck too small to be seen under the most powerful microscope. Do the same with the entire known universe and what you wind up with is an incredibly heavy English walnut— maybe less.

Not that you could actually do such a Big Squeeze, even if you had a strong enough vise. Long before you got the Universe down to walnut size its gravity would rip a hole in the fabric of the cosmos and send everything into a gigantic black hole. But the thought experiment remains. Solid matter is anything but solid. It's composed of atoms, which are themselves composed of subatomic particles and a lot of empty space. Inside the subatomic particles are smaller-yet particles and even more empty space—the whole works kept from collapsing into itself by the balancing acts of kaleidoscopic motion and attractive and repulsive forces. Such is the picture produced by what is now referred to as the Standard Model of the atom.

More than 2,400 years ago, a Greek philosopher named Democritus (~460-370 B.C.E.) wondered what would happen if you took a lump of—well, anything—and divided it in half again and again and again, *ad nasueum*. His conclusion: eventually you would reach the smallest possible bit of whatever it was you started with, beyond which you could subdivide no farther. That bit he dubbed the atom (or *atomos*), which is Greek for *indivisible*.

It would be 2,300 years until Albert Einstein actually found incontrovertible evidence that atoms exist, proving Democritus's insight to be correct—and by then physicists were already trying to pry Democritus's atoms into smaller pieces: the now-familiar electrons, nucleus, protons, and neutrons. But today's physicists still believe the same thing Democritus preached: that if you delve deeply enough into matter, you will indeed find something *atomos*…something indivisible. But the deeper we delve, the less sure we are of what it might be.

Democritus' insight wasn't the result of science as we know it today. It was a thought experiment, and a purely philosophical one at that. The question that inspired him and an entire group of Greek "atomists" was why every form of matter we know hadn't long ago degenerated into some kind of sludge in which everything was hopelessly intermixed. The only answer, they believed, was that it was composed of particles that could be alternately mixed into the Greek equivalent of "grey goo" and reassembled into rocks, plants, animals, you, and me.

The atomists argued that these atoms must be separated by void spaces, because otherwise everything in the world would be locked together. Ice would be rock and you and I would be statues. Furthermore, they concluded, different atoms had to have different shapes. Atoms of iron had to be strong and solid, with some kind of hook-like structure to lock them into swords, plowshares, or Golden Gate bridges. Water atoms had to be smooth and slippery. Salt atoms (because of their taste) had to be sharp and pointed, while air atoms must be light and whirling.

Plato carried the theory even further, replacing the hooks and points with geometrical shapes: tetrahedrons, cubes, octahedrons, and icosahedrons. These, he believed, represented the atoms of fire, earth, air, and water, with the shapes related to how easily they flowed or moved. Earth was a cube because it stacked firmly. Water was an icosahedron whose numerous sides let the atoms slide and tumble more easily. He was also drawn to this idea by the fact that each of these shapes could be decomposed into triangles that could, in turn, be reassembled into the other polyhedrons. Modern subatomic theory this was not, but it was the first suggestion atoms might be composed of even smaller pieces.

British philosopher Bertrand Russell (1872-1970) called it all a lucky guess. Others have noted that the Greek atoms sound more like molecules than today's atoms. Even in the Greek world, it was a relatively short-lived theory, largely thanks to Aristotle (384–322 B.C.E.). "Aristotle tended to be skeptical of ultimate entities," says Gregory Goode, a historian of science at the American Institute of Physics in College Park, Maryland. "He would have said Democritus and [other atomists] were too speculative." In particular, Aristotle didn't like Democritus' theory because it required a void between atoms (other than Plato's tightly stacking "earth" cubes). Such a void could not exist, he argued in one of those mind-bending twists of Greek logic, because the moment you identified a part of the universe as "void" you acknowledged it was actually "something" and therefore not "nothing."

However hard that line of reasoning is for modern minds to parse, Aristotle's influence pretty much settled the debate for more than 2,000 years. There were a few contrarians, such as Persian scholar Abū Hāmid Muhammad ibn Muhammad al-Ghazālī (A.D. 1058-1111) or medieval Frenchman Nichols of Autrecourt (A.D. ~1299-1369), the latter of whom also believed time and space were composed of atom-like instants and points. But these were the exceptions. Basically, says Axel R. Reisinger, translator of *The Atom in the History of Human Thought,* by French quantum chemist Bernard Pullman (1919-1996), "there were precious few milestones between the Greeks and the 17th Century. There was a huge gap when nothing much happened."

One 17th Century scholar who helped resurrect atom theory was Galileo. Writing in 1623, he suggested that the sense of taste came from "minute particles" in food that affected the tongue "according to the various shapes, numbers and speeds of the particles." The same, he added, might explain the sense of smell. Again, this sounds more like molecular theory, but again, it's intriguingly on the right path.

The next 280 years saw ever-more-refined attempts to answer the same basic question. Descartes (1596-1650) shared Aristotle's dislike of the void, believing all space must be filled with matter. Newton (1642-1727) wasn't sure about atoms but had no trouble with the void. "He thought the active parts of matter were very, very small," says Goode, "and what was really important were their interactions over space." Then in approximately 1803, English chemist John Dalton (1766-1844) published what would become the first accurate list of the elements. Initially he had only six: hydrogen, oxygen, nitrogen, carbon, sulfur, and phosphorus. But he had a very clear idea that these were composed of indivisible atoms. "We might as well attempt to introduce a new planet into the solar system, or to annihilate one already in existence, as to create or destroy a particle of hydrogen," he wrote in 1808.

By the 1890s, atomic theory was well established, largely thanks to chemists who'd been working out the chemical formulas and the structures of increasingly complex molecules. But it took Einstein to actually measure the mass of an atom.

He did it by looking at the odd vibrations of dust motes. As far back as 1827, Scottish biologist Robert Brown had noticed that water-suspended pollen grains, viewed under a microscope, had a tendency to jitter randomly. At first he thought he'd found some new kind of life force, but when he found the same vibrations in inorganic matter, he was merely perplexed. By Einstein's time, it had been long hypothesized that the pollen fragments, etc., were being shoved around by random collisions with water molecules.[44] But atoms are small and pollen fragments comparatively large. Why was it that the myriads of collisions didn't balance out, leaving the pollen grain motionless?

In a 1905 paper published only months before his seminal pieces on special relativity and $E=mc^2$, Einstein found the answer, demonstrating that the particles were being moved by the slightly unbalanced effects of trillions of collisions in a process mathematically akin to that which allows fluids to move across cell membranes via osmosis. Almost as a side-effect of this theory, he could calculate the size and weight of individual water molecules, already known to be composed of H_2O. At long last, he had the weight of hydrogen, and by easy chemical calculations, everything else.

But ironically, it was already known that the atom was not *atomos*. Nearly a decade before, British physicist J.J. Thomson had discovered the electron, and correctly deduced it was subatomic. Thomson did it by using vacuum tubes to study "cathode rays," mysterious emissions that occurred when high-voltage current was shot through a vacuum, much as in an old-fashioned TV tube. First, he put a metal slit inside a tube and used a magnet to scan the beam across the slit. From the way the beam passed through the gap, he concluded it was composed of particles, not waves. Then he tried to bend the beam with an electric field, discovering that the particles had an enormously high charge-to-mass ratio—a thousand times higher than that of ionized hydrogen. They were either incredibly tiny or incredibly highly charged.

---

[44] Today, this movement is called Brownian motion.

As it turns out they were incredibly tiny. Subsequent experiments, the most famous of which was the Millikan oil drop experiment, done in 1909, used various tricks to isolate single electrons and measure their charge. Millikan's brainstorm—today replicable in any high-school physics class—was to impart a tiny static charge to oil droplets by spraying them out of a nozzle much like a perfume sprayer. If he was careful, he could make sure that most droplets picked up no more than a few electrons' worth of charge. Then, by using an electric field to hold the droplets stationary against gravity, he could calculate their individual charges. The jumps between two, three, or four electron-units allowed him to calculate the charge on individual electrons.

In the course of barely a decade, scientists had gone from being unable to measure anything about atoms to knowing not only their masses, but the mass and charge of the electron. It was the first step in what would become an explosion in atomic knowledge. But something was still missing. If atoms included negatively charged particles, there had to be a balancing positive charge. Today we know it's the proton, but the scientists of the time saw the atom as something akin to a blob of chocolate chip cookie dough (positively charged), with the electrons embedded in it like negatively charged chips. (Other scientists referred to this as the plum-pudding model or the raisin-cake model, depending on their preferred desserts.)

The test came in 1909, the same year Millikan was nailing down the charge on the electron, when New Zealander Ernest Rutherford decided to see if he could probe the atom by hitting it with high-speed sub-atomic "bullets." He chose alpha particles, which were a form of radiation he'd discovered a decade earlier. These particles, he knew, carried twice the charge of an electron, with a mass equivalent to the helium atom.

Using either radium or polonium-214 as a source (accounts vary), he beamed high-speed alpha particles at a sheet of ultra-thin gold foil, 0.0005 mm thick. (He chose gold presumably because it was easy to work with and had big atoms, known to be about 50 times heavier than alpha particles.) Under the plumb-pudding model, the alpha particles should have passed right through the gold sheet, veering slightly if they came close enough to an electron for their charges to interact. But that's not what happened. Most of the alpha particles behaved as expected. But a few hit the foil . . . and rebounded.

Rutherford could hardly have been more surprised. But he wasn't a man noted for being long at a loss for words ("an alleged scientific discovery has no merit unless it can be explained to a barmaid," he once said), and soon enough came up with a suitable way to describe his reaction: "It was almost as incredible as if you fired a 15-inch shell at a piece of tissue paper and it came back and hit you."

What he'd discovered was that there was something a lot more solid in the plum pudding than simply the tiny electron-plums—something we now know to be the nucleus. But he'd also discovered that the nucleus was tiny—otherwise, a lot more of his alpha particles would have bounced back at him. What this meant was that other than the electrons (which are very tiny), most of the atom was empty space. In other words, Democritus's void existed not only *between* atoms, but *within* them as well.

Rutherford had also discovered what remains the primary experimental tool of sub-atomic physics: using high-speed particles as bullets to be smashed into atoms to see what types of bits might come flying out. In a string of experiments conducted in 1917-1919, he also discovered that alpha particles, passing through nitrogen gas, left the unmistakable signs of fast-moving hydrogen nuclei. These hydrogen nuclei (which he later dubbed protons), he realized, were being knocked out of the nitrogen nuclei by the impact of the alpha particles.

Then, in 1932, British researcher James Chadwick published a brief paper in *Nature* called "Possible Existence of a Neutron," in which he reported an uncharged particle with a mass similar to the proton, produced when beryllium was bombarded with alpha particles. And with that discovery, our basic vision of the atom was complete. Atoms, it seemed, weren't truly *atomos* but actually consisted of three well-defined particles: protons and neutrons in the nucleus, with electrons spinning around them like planets around a sun.

Not that progress was actually quite that tidy, because it took the scientists of the time a while to realize the import of what they were finding. "We love these milestones: the electron was discovered in 1897, the nucleus in 1911," says Goode. But it wasn't that straightforward. "Rutherford, at first, didn't think he had done anything important. Nobody else did, either. The reaction in 1911 was basically silence."

Chadwick, Goode adds, once said that it's no good writing history in hindsight. "You have to write history as if you are looking over the shoulders of these brilliant scientists who are mightily confused. Sometimes they weren't sure what they had until quite a long time after they had it."

Meanwhile, as far back as Rutherford's discovery of the nucleus, the plum-pudding model of the atom had quickly given way to the "planetary" or "solar system" model in which electrons circled the nucleus in orbitals that didn't behave quite like planets in a solar system[45] but at least seemed similar enough to provide a tidy analogy that would pass Rutherford's "barmaid" test.

And then, chaos.

Using Rutherford's atom-smashing methods at ever-higher energies, physicists in the 1940s and 50s began blasting a whole zoo of subatomic particles out of nuclei—hundreds of different types of particles called baryons and mesons that were neither protons, neutrons, nor electrons. "There are an outrageous number of them," says Don Lincoln, a particle physicist at the Fermi National Accelerator Laboratory in Batavia, Illinois.

In the 1960s and '70s all of this led to what is now known as the Standard Model in which everything is composed of things called leptons, quarks, and gauge bosons. There are more than a dozen of these (not counting their antimatter equivalents), ranging from the photon to six different types of electron-like particles (some with mass, some without). There's also the recently discovered Higgs boson, which defies any attempt to apply Rutherford's "barmaid" rule in explaining it but plays a critical role in keeping the Universe from being utterly massless.

But in many ways the quarks are the most interesting, because they allow us to peel back yet another layer and peer

---

[45] The differences involved the fact that orbitals seemed to have only certain allowed energy states, and that electrons sometimes behaved as waves instead of particles. But electrons are the real of chemistry, and the topic of this chapter is the atom.

inside the proton and the neutron: particles previously believed to be indivisible.

Quarks are to protons and neutrons what electrons and the nucleus are to atoms: incredibly tiny subcomponents that exist largely in a void. Just as the subcomponents of atoms appear to compose only one trillionth of their volume, quarks appear to comprise only a trillionth or less of the volume of protons and neutrons. Once again, when you peer deeply, what you find is mostly the void.

Quarks were initially believed to come in two basic types, somewhat arbitrarily named "up" and "down". (The names actually come from 1950s terminology relating to spin states.) "Up" quarks have a charge of +2/3. "Down" quarks are -1/3. Assemble them in the right combination and you can make a proton: up-up-down. For a neutron: down-down-up. Initially, quarks existed merely in theory, but as ever-more-powerful atom smashers were developed, both up and down quarks were discovered in 1968.

Then, scientists found that both types of quarks actually came in three flavors. Also carrying a charge of +2/3, but distinguishable from up quarks on the basis of mass were the "charm" and "top" quarks. Analogous to the down quark were the "strange" and "bottom" quarks. Why the peculiar names? Blame it on physicists with too much time on their hands, says Lincoln. "[They're] being cute." It also took a long time to find all of the quarks, Lincoln says, noting that it was 18 years from the discovery of the bottom quark (the fifth to be identified) in 1977 to the ultimate find of the top quark in 1995. "We looked and looked," he says.

But why are there so many types of quarks when only two

are needed to make protons and electrons?

"The short answer is nobody has a clue," Lincoln says.

Another oddity was the mass quarks proved to have when they were discovered. "On the face of it, you would think quarks would be about one-third the mass of the proton," Lincoln says. "That makes perfectly good sense—and is absolutely and completely wrong." Rather, he says, the up and down quarks add up to only about 2 percent of the mass of the protons and neutrons they combine to form. "Protons and neutrons aren't like three marbles stuck together," he says. "The way to think of the proton and neutron is kind of like a lottery machine where you have three little flecks of Styrofoam, and the wind is blowing them around. The flecks are the quarks. The wind is the strong nuclear force, which is the thing that holds [them] together." The missing 98 percent of the mass is the Einsteinian mass-equivalent of the energy represented by that force.

Weirder yet, some of the other quarks are quite heavy—the top quark, for example, is heavier than most atoms, roughly comparable to the metal tungsten. Not surprisingly, such quarks don't play much of a role in our ordinary world. "At the very beginning of the Universe, they winked into existence [then decayed and vanished]," Lincoln says. "Now they show up only in cosmic rays and our accelerators."

Are quarks the ultimate building blocks? "Nobody knows," says Lincoln. But the fact there are three types of each of the basic types of quarks is "suggestive" that something deeper may be going on. Scientists already have a name for that next level: preons. But so far nobody has found it. In fact, nobody has yet determined the size of quarks themselves. "All we know is that if they had a size [larger than] 10^{-20} meters, we'd have seen it," Lincoln says. "Our equipment can see on the order of 10^{-20} meters—five orders of magnitude smaller than the proton—and even though we've looked very hard, with the precision of our equipment they still look very pointlike."

But that doesn't mean preons don't exist. "Personally, if I were a betting man, I would think there probably are constituents inside of quarks," Lincoln says. And he notes, similar particles might exist within the electron. "There is zero evidence, but that's the idea."

And beyond that? Yet again, the answer is who knows.

"There could be many levels of matter," Lincoln says. "One hypothesis for the ultimate end of the game is that at the smallest scale are little strings—not pointlike, but little sticks of spaghetti or hula hoops—and depending on how they vibrate they are the different particles."

How many levels of sub-sub-sub particles might lie between what we know and these strings (sometimes called superstrings)? Well, atoms are on the order of 10^{-10} meters in diameter. Protons are 10^{-15}. Quarks are 10^{-20} meters or smaller. "So we're talking about orders of magnitude of about 100,000 between scales," Lincoln says. If superstrings are real, he says, they are on the order of 10^{-43} meters in length—a size below which "physics as we know it will break apart—which suggests that this is the building block. But that's a size scale totally beyond anything we can see in our lifetimes."

But however many levels actually exist, we know one thing for sure. "At the deepest level we understand it," Lincoln says, "the atom is mostly empty space [composed of] things that have almost no size, and the force fields that hold them together."

Author's Note

When I was in college, I was in a program that required me to take an "honors" elective each quarter. The topic didn't matter; it just had to be something wildly outside my major field. My near minor in history stems in part from that program.

The final quarter of my senior year, I did something even more exotic: I took an upper-level course in economic geology, taught by a professor named…I kid you not…Stonehouse. In geology, "economic" means mining and oil and gas, so despite it's upper-level status, the course was accessible to non-majors. And Stoney, as we knew him, was accustomed to having honors students from other departments, so he designed his course accordingly.

Late in the semester, we took a three-day field trip into Canada. Non-majors weren't required to go, and I may have been the only one to take the opportunity. We drove from East Lansing, Michigan to the north shore of Lake Huron, somewhere near Sudbury, Ontario, where we toured a nickel mine and a uranium mine. But what I really remember was a morning of road-cutting.

Geologists are probably the only people in the world other than highway contractors for whom the word "roadcut" can be a verb. In geology, it means touring the highways, looking for cuts that expose interesting rocks. As a non-major I wasn't required to join this expedition, but I wasn't about to pass up the chance to explore. At one roadcut, however, the excited technical discussion went way over my head, so, after making sure the rest of the group would be there for at least a half-hour, I scrambled to the top of the nearby hill, where I found myself overlooking a north-country wonderland of granite, spruce, pine, and hemlock, with the roar of whitewater somewhere in the distance. I lingered as long as I dared, as it soaked in on me that I might have missed a possible career. Astronomers have the stars. Geologists get to hike into places like this.

Since then, I've become a fairly good amateur geologist. I've written dozens of articles about geology. Which brings us to our next suite of stories. Before history lies "prehistory." And before that, lies paleohistory. The next three stories work backward millions, then hundreds of millions, and ultimately billions of years into Earth's past.

Stoney died in 2011 at the age of 89. These are for him.

THE PREHISTORY OF GLOBAL CLIMATE CHANGE

In 1994, a research team led by Kirk Johnson, curator of paleontology at the Denver Museum of Nature and Science, discovered the fossil remains of a 64-million-year-old forest in a roadcut near Denver, Colorado. Since then, the scientists have identified 170 species of plants in this forest, some with leaves up to twenty-four inches long. Many of the leaves had long, narrow tips—called *drip tips* by biologists. Such leaves shed moisture more easily after a rainfall, allowing them to drip-dry between showers so they don't provide moist surfaces for moss or parasites. They're common in wet climates, but not in modern-day Colorado, where rainfall is far too scarce for such leaves to be necessary. In fact, they're only found in rainforests receiving 100 or more inches of precipitation per year. Johnson's conclusion, published the June 28, 2002 issue of *Science*: 64

million years ago, Colorado was much warmer and wetter than it is today.[46] Nor was this climate change simply due to the vagaries of continental drift, which can, over the course of millions of years, move a landmass from the tropics to higher, drier latitudes. Sixty-four million years ago, Colorado was located at very close to the same latitude as it is today.

The study of ancient ecosystems is called paleoecology. And in today's world, when global climate change is an international concern, it is of more than passing interest. By studying how plants and animals responded to prior climate upheavals, paleoecologists hope to shed light on what might be in store for us in years to come.

Table 1. Geological Timeline

Era	Period	Epoch	(Mya)[a]	Highlights
Cenozoic	Quaternary	Holocene	0.01	Civilization as we
		Pleistocene	1.8	know it.
	Tertiary	Pliocene	6	Ice ages.
		Miocene	24	
		Oligocene	37	
		Eocene	58	
		Paleocene	65	
				Rise of mammals.
Mesozoic	Cretaceous	Each period is divided into numerous epochs	135	K-T extinction. Big dinosaurs.
	Jurassic		208	
	Triassic		248	Small dinosaurs.

---

[46] Kirk R. Johnson and Beth Ellis, *Science*, 28 June 2002, pp. 2379-2383.

Paleozoic	Permian		290	Permian extinction.
	Pennsylvanian		320	Carboniferous swamps. Ice ages.
	Mississippian		355	
	Devonian		410	Age of Fish.
	Silurian		438	Land plants.
	Ordovician		505	
	Cambrian		545	Trilobites.
Precambrian	Ediacaran		600	Filter-feeding animal communities on seafloor.
	Proterozoic Eon[b]		2,500	Ice ages may have reached equator, creating "Snowball Earth."[b] Photosynthesis adds oxygen to atmosphere.
	Archean Eon		3,900	Earliest life.
	Hadean Eon		4,500	Earth formed. Not a friendly place.

[a] Precise dates of some periods are open to debate. Mya means "million years ago."

[b] Eons are bigger divisions than periods, but not as big as eras.

Big Nasty Rocks, Volcanoes, and Thermal Maxima

The world has seen many climate changes, but the most famous occurred 1.4 million years before Johnson's rainforest, marking the division between the Cretaceous and Tertiary periods of geological history (see Table 1). Called the K-T boundary, it's best known as the end of the age of dinosaurs. But dinosaurs weren't the only species affected. In a geologically short period of time, something on the order of 70 percent of the Earth's species went extinct.

Table 2: Ancient Climate Changes

	Permian Extinction	K-T Event	PETM	IETM
Date	250 Mya	66 Mya	56-43 Mya	52 Mya
Probable cause	One-two punch of uncertain origin. Volcanism or an asteroid strike?	Global warming followed by a big, nasty rock from space	Global warming spanning several million years	Rapid spike in global temperature due to greenhouse warming from methane hydrate release. Gradual return to baseline
Short-term effects	Extinction of 90% of ocean species; destructtion of woodlands	Massive extinctions	Species survival	Increased species diversity due to intercontinental migrations; few extinctions
Long-term effects	Woodland recovery delayed 4-5 million years.	Species diversity reduced. Opened door for age of mammals.	Warming associated with increased species diversity	New animals dominate North America. Little impact on plant communities.

146

But the K-T boundary isn't the only major climate change that occurred between 250 and 43 million years ago (see Table 2). There were, in fact, three others: the Permian Extinction, the Paleocene-Eocene Thermal Maximum (PETM), and the Initial Eocene Thermal Maximum (IETM). Let's examine the others in historical order.

The Permian extinction was the most massive die-off in the history of the Earth, but because it didn't involve dinosaurs, it's not as well known as the K-T extinction. Its cause is subject to debate. It is known, for example, that this was an era of extraordinary volcanic activity in Siberia. This volcanism (which has no parallel in human history) would have thrown massive volumes of ash into the air—enough to considerably impact the Earth's climate. Another hypothesis is that the Earth was hit by an asteroid. Yet another is that this was about the time most of the Earth's landmasses came together to form the supercontinent Pangaea, and perhaps something about having all of that landmass combined did something to reduce biodiversity. (What, exactly, isn't well-articulated.) Or perhaps the cause was something more prosaic. Whatever happened, paleoecologists estimate that 90 percent of marine species and a good fraction of terrestrial species died.

Cindy Looy, a plant ecologist at the University of California, Berkeley, has examined fossilized pollen and plant spores trapped in sediments laid down before and after the Permian event and found that before the Permian extinction, the world was dominated by conifer and seed-fern woodlands. Afterward, the woodlands disappeared—so thoroughly that it took four to five million years for them to reestablish. But, interestingly, the collapse wasn't immediate: instead, it stretched out over perhaps

200,000 years. This may indicate that the Permian ecosystem received a one-two punch. Perhaps the Siberian volcanoes weakened the ecology enough for a subsequent asteroid strike to clobber it more thoroughly.

The K-T event also appears to have involved two distinct causes. For about a half-million years, the Earth had been warming—rapidly by geological standards. Johnson and Peter Wilf of Pennsylvania State University have studied more than 22,000 plant fossils from this era, looking for last appearances of each species.[47] A significant fraction of those that went extinct appear to have done so slightly before the K-T boundary, during the period of rapid-but-not-catastrophic climate change. Then something happened and a great many more species went extinct virtually overnight.

Prevailing theory, of course, is that the coup de grace was the fabled "dino-killing" asteroid strike in Mexico's Yucatan Peninsula. And as in the Permian, the Earth was slow to recover: terrestrial ecosystems remained impoverished (from a species-diversity perspective) for millions of years after the asteroid impact.

The Permian and K-T extinctions give us our first paleoecology lesson, and it's a sobering one: if you damage the Earth's ecosystems badly enough, it takes a very, very long time to recover. The PETM offers a different lesson. Its "thermal maximum" was the result of a steady warming trend extended over an interval of several million years—roughly from 55 to 43

---

[47] Peter Wilf & Kirk R. Johnson, *Paleobiology*, 30 (3), 347 (2004).

million years ago.[48] It produced warm climates, worldwide, but wasn't associated with any abrupt incident, such as a dino-killer rock from outer space. Nor was it a period of mass extinction. Most plant species sailed through it—changed in range and relative abundance, but not wiped out. The message here: ecosystems adapt much better to slow climate changes than fast ones.

Unfortunately, this isn't as good news as we might hope for our own future, because from a geological perspective, human activity has been altering ecosystems nearly as suddenly as a meteor strike. Not to mention that in addition to global warming, we endanger them in much more prosaic ways, such as habitat destruction, pollution, and fragmentation by highways, farms, and parking lots. Even our penchant for plowing farmlands or bulldozing subdivisions for home construction may play a role: in a 2003 study, Ron Amundson, a soil scientist at the University of California, Berkeley, cataloged rare and "endangered" soil types and found that they were linked to endangered plant species. Some plants, in other words, require very, very specific types of soils, and by moving dirt around we are contributing to their destruction in yet another previously unsuspected way.[49] Overall, the conclusion of most scientists is depressingly consistent: the current, human-caused extinction could easily rival the K-T boundary...if not the Permian Extinction. "We could easily be looking at mass extinction of

---

[48] One way to determine ancient temperatures is from the ratios of oxygen isotopes in limestone deposits. At different temperatures, the seashells that formed these rocks contained different levels of the relatively uncommon oxygen-18 isotope.

[49] Amundson, et al, *Ecosystems*, 24 July 2003.

half of our plant species over the next 100 years," Wilf says.

If you're looking for hope, it might lie in the IETM, rather than the PETM.

The ITEM was a spike of super-warm temperatures embedded in the middle of the PETM, corresponding to an additional warming of between 4°C and 8°C at high latitudes (7°F to 14°F) over the course of no more than 10,000 to 20,000 years.[50] Most likely, it was triggered by a massive release of methane hydrates from the seafloor.

One way this could happen would be if the slower warming of the PETM reached a critical level that allowed these substances' release. Or maybe a comet strike or big earthquake disturbed a large part of the ocean bottom. Whatever started the process, once the hydrates—an odd form of ice that contains methane as well as water—escaped from the seabed they would have released large amounts of methane (a powerful greenhouse warming gas) into the atmosphere.[51]

The result would have been a dramatic short-run change in climate—though again "short-term" to geologists isn't the same as it is to the average person: in this case, it took "only" 80,000 to 100,000 years for the climate to return to normal PETM levels. Meanwhile, the IETM warming had allowed a dramatic change in the mammalian species present in North America.

---

[50] The transition might, in fact, have been even more abrupt, but are limits to geologists' abilities to resolve climate change in rocks laid down that long ago.

[51] Evidence for this comes from carbon isotopes in rocks formed during that period, a marker of the source of the carbon from which these rocks were formed. Methane hydrates still reside on the ocean floor, and a sufficiently large climate disturbance could cause large quantities of them to rise to the surface.

Sixteen new genera of mammals appeared in the West, coming to represent half of the total mammalian population. These included ancestral pigs, dawn horses, and primates.[52] Most likely, Wing says, these animals migrated to North America from Asia, via polar land bridges previously blocked by harsh winters.

Normally, invasive species are thought of as destructive—crowding out the natives or out-competing them for crucial resources, just as rabbits and wild dogs have wreaked havoc in Australia. But that's not always the case. In the Eocene, the pigs, horses, and other large animals merely added to North America's existing ecosystems, increasing their diversity.

Interestingly, plant species did not show a similar spread. A few reached North America from Asia, but most never became abundant. Why the difference? One reason is that plants also had to migrate across high-latitude land bridges—slowly seeding and reseeding themselves along the route. And even if the poles had warmed considerably, these plants would still have had to contend with the long, dark winters. For many, that itself may have been a major barrier.

Overall, there are four important lessons that can be drawn from the IETM: (1) Greenhouse-warmed climates can change rapidly. "That's sobering," Wing says. (2) Sometimes, ecosystems can adapt to fairly rapid change (though it should be noted that even the IETM wasn't really all that rapid by human standards). (3) Animal species can change their ranges rapidly in response to climate change, with permanent alterations in the resulting ecologies. (4) Plant migrations aren't always as rapid.

---

[52] Many of these later went extinct, but that wasn't until much more recently.

Yo-yo Seas . . .

Let's roll the clock further back to the Pennsylvanian Period of the late Paleozoic. It was the era, 300 million years ago, in which coal seams (including those of—guess where?— Pennsylvania) were formed from vegetation buried in swamps: an era when both America and Europe lay near the equator. It was also an ice age, and as the ice caps waxed and waned, sea levels fell and rose, causing the ocean to cyclically retreat and advance from the coastal plains in which the coal was being formed. This had two effects. First, the shifting sea alternately wiped out the swamp ecosystems and exposed ocean bed on which new ecosystems had to develop from scratch. Secondly, the cyclical infusions of salt water percolated into the coal beds and preserved detailed cellular structures of the buried plants. The result is a bonanza for paleoecologists. It's even possible, says Wing, who with his colleague Bill DiMichele has studied many of these coal beds, to tally what fraction of the ecosystem's biomass was comprised by each species.

The overall picture is one that's been known for many years. The swamps were woodland ecosystems comprised of a more alien-looking mix of vegetation than that found in many science fiction stories: landscapes dominated by giant clubmosses, enormous horsetails, conifers, seed ferns, and tree ferns.

But for paleoecologists, the most interesting observation is that with each advance and retreat of the ocean, the same groups of species returned, forming very similar ecosystems, again and again.[53] Apparently, the swamp species were simply shoved to

---

[53] W.A. DiMichele, et al, *Ann. Rev. Ecology, Evolution and Systematics*, 35:285-322 (2004).

new, inland habitat from which they could revegetate the coastal plains once the oceans withdrew. It's a reminder that while individual plants can't get up and walk away from danger, plant species can save themselves by reseeding to new, now-suitable habitat. This offers some reassurance for our own future. Humans, after all, are very good at moving plant and animal populations around. If we drive species out of their current environments, we'll do our best to establish them in new environments (presuming, of course, that suitable environments exist).

It should be noted, though, that the coal swamps of the Pennsylvanian Period were comprised of species that could spread easily enough not to be wiped out by the ocean. (Those that couldn't would have been eliminated by its first advance.) Not all plant communities are so easily mobile, even when humans try to help. A collection of mini-ecosystems near Merced, California offers a disturbing case in point. Considered by ecologists to be one of the world's most important biodiversity "hotspots," the Merced region is a mosaic of river channels, ridges, basins, and vernal pools. Each contains a unique soil, deeply enough linked to individual plant communities that people working to preserve endangered species have found it difficult to transplant them, even to closely related soils.[54]

Near its end, however, the Pennsylvanian Period went through a transition from a sub-period known as the Westphalian to one called the Stephanian.

During the Westphalian, the coal seams revealed 48 plant

---

[54] Amundson, Guo, & Gong, *ibid.*

species, of which 30 were trees and 18 were groundcover. It was this collection of plants that persisted for millions of years, through each cycle of ocean advance and retreat. But something happened at the transition from the Westphalian to the Stephanian. The species composition suddenly shifted to 17 types of trees and 25 types of groundcover. Even more significantly, more than half of the species in the Westphalian swamps weren't among those in the Stephanian swamps. What happened?

Apparently, Wing says, the tropics had become drier, perhaps developing a longer, harsher dry season than was seen during the Westphalian. One theory is that this was caused by a change in the South Polar ice cap. That would have altered the flow of moisture, which in a phenomenon called *inter-tropical convergence* is pinched away from the ice caps toward lower latitudes. The change in the South Polar ice cap may have been short lived, but by the time the tropics regained their normal moisture, the new plant species had become well established and the old ecosystems never returned. The lesson is the converse of that from the IETM, and yet another warning for humans: a geologically short-lived climate change can cause irreversible ecological change.

. . . and a Sheltered African Valley

Irreversible ecological changes aren't necessarily bad for humans. In fact, a similar climate shift may have spurred the emergence of our ancestors. Let's set the paleoecological way-back machine on fast forward and zoom from the Pennsylvanian to the African Rift Valley between 3.5 and 1.5 million years ago. At the start of this era, the Rift Valley was largely comprised of canopied forest. Early hominids were present but genus *Homo* had yet to appear, and *Homo sapiens* was still far in the future. But by the end of this era, much of East Africa had dried to a brushy grassland, and genus *Homo* was firmly established.

Anna K. Behrensmeyer of the National Museum of Natural History has studied the link between these two changes. Her research centers on the Gulf of Aden and the Omo River Valley of southern Ethiopia.

The Gulf is an arm of the Red Sea that collects wind-blown dust from the Sahara Desert. Core-drilling allows sedimentologists to determine the climate history of the Sahara, both by the quantity of dust and by the isotope ratios of oxygen trapped in the sediments, which are an indicator of the temperature at which they were deposited. These core samples reveal that at the start of this era, northern Africa's climate was warm and stable. Then, it became cooler and more variable.

In the Omo Valley, paleontologists have recovered more than 1,400 hominid specimens—enough to allow a pretty good determination of which hominids were present at what times. The fossil record also shows much about the other plant and animal communities populating the valley, and digs have been conducted at enough sites to allow researchers to map the ecologies represented by each of the region's sedimentary strata.

The valley is also a paleontologist's paradise, with many layers of floodplain deposits interleaved with layers of volcanic ash. That ash can be dated by two methods: the decay of radioactive elements contained in it and by paleomagnetism, which detects the imprint of the Earth's magnetic field in rock strata, using changes in that field to determine the era in which the rocks were formed. In combination, these two allow an unusually detailed assessment of the valley's ecological history.

Through all of this time, the Omo Valley was an area teeming with wildlife: antelope, giant pigs, monkeys, and more. Overall, Behrensmeyer's team has divided it into four broad ecosystems: canopied forest, open woodland, wet grassland, and dry brushy grassland. Pigs and monkeys thrived in the dense forest; different pigs, monkeys, and an extinct form of kudu lived in the more open woodlands. The wet grasslands drew buffalo and waterbuck. The brushy grasslands drew antelope, pigs, baboons, and wildebeest.

Not surprisingly the extent of these ecosystems changed as the climate shifted. For the first million years, from 3.5 to 2.5 million years ago, the change was gradual, producing a slow shift from canopied forest to open forest, wet grassland, and eventually to dry grassland. But the Omo Valley was sufficiently well protected from the climate changes afflicting the rest of North Africa that all four ecosystems continued to exist— preserving substantial tracts of canopied forest, which provided habitat for forest-dwelling animals, including protohumans.

Then, about 2.5 million years ago, the valley's ability to buffer the effects of neighboring climate changes was exceeded, and its own climate became more unstable—subject to erratic swings vaguely comparable to the repeated ingress and egress of seas in those Pennsylvanian-era coastal plains. The shift from stability to variability occurred quite rapidly, but there still wasn't a complete elimination of the canopied forest. Remnants persisted, providing refugia in which the forest dwellers could continue to exist while the land around them changed.

These climate swings put pressure on the protohumans, but the refugia gave them hundreds of thousands of years to adapt.

And major adaptations were needed, because the increasingly open habitats were much more dangerous places to forage than the canopied forests. It was during this time that the first members of genus *Homo* began to appear, along with their first artifacts. Shortly after that, *Australopithecus*, a human precursor that had previously been quite widespread, disappeared.

Perhaps, Behrensmeyer says, it was climate change occurring at just the right pace that led these early hominids to develop into cultural beings—beings who could work together for defense and food-gathering, beings who could manufacture tools and teach these skills to their children, beings who would eventually erupt out of Africa to colonize the globe . . . and eventually worry about changing its climate.

Not-so-paleo- Ecology

Beginning in about 1910, the American West experienced a string of unusually wet years, affecting everything from California's Mojave Desert to the sagebrush steppes of eastern Oregon and Washington. The Great American Desert wasn't exactly squelchy, but there was enough water for farming, and homesteads and agricultural communities sprouted everywhere, like spring mushrooms.[55] Then the climate reversed, not merely returning to normal but overshooting into the Dust Bowl drought of the Great Depression. The farmers disappeared like tumbleweeds on the desert wind and dust storms blew topsoil as far away as the Atlantic Ocean. Okies lined the highways to California and John Steinbeck wrote *The Grapes of Wrath*.

---

[55] I know this from decades of exploring ruins in the area, but for specific data, see, Connie A. Woodhouse, et al., "The twentieth century pluvial in the western United States," *Geophysical Research Letters*, 32, L07701 (2005).

Can all of that happen again?

In the deep paleoecology we've been discussing so far, such changes occur too quickly to show up in the fossil record. But there's another branch of paleoecology that examines more recent climate history, and it has the ability to resolve changes that occur over human-scale time intervals. It can't tell us whether the Dust Bowl might repeat, but it does give us the disturbing news that even by fairly recent standards, the Dust Bowl was a *minor* drought.

One way of studying prehistoric climates is by looking at the growth rings on trees. The oldest in North America are bristlecone pines dating back about 5,000 years, but they live on isolated mountaintops in Nevada, Utah, and southeastern California. For broader climate records, you need trees that are more widely dispersed, even if they're not as old. One good candidate is the juniper, which can live 750 or so years and is common throughout the West. These and other old trees indicate that shortly before the first Europeans reached North America, the continent went through a drought considerably more severe than anything in recorded history. It lasted for several decades and stretched from Central Mexico to Montana, from the Rockies to Virginia. And, looking back to the Fourteenth and Fifteenth centuries, scientists have seen even longer and deeper droughts. The Dust Bowl was economically devastating, but a repeat of any of these prior droughts would be far worse.

Tree-ring studies are only one way of getting a detailed view of prehistoric climates. Another is by studying pollen grains trapped in bog and lakebed sediments. In eastern Michigan, for example, there are bogs that have existed for 3,500 years that show an interlude in which there was a sudden switch in tree pollen from beech to pine, plus an increase in the amount of charred wood—a sign of forest fires. In combination, these indicate that at one point eastern Michigan dried up swiftly (over the course of only a few decades) then stayed dry for 1,000 years. [56] By human standards, that's more than just a drought: it's a long-term climate change, and not one most of us would see as being for the better.

Scientifically, the bog studies are important because they have helped change old theories of plant succession, which posited a gradual change in plant communities, as each ecosystem slowly prepared the way for its successor. Supplemented by insights gained from watching the recovery of Mt. St. Helens after its cataclysmic 1980 eruption, the bog studies support a new theory of ecological change, which says that ecosystems shift, not gradually, but in staccato jolts that occur when something overwhelms an old ecosystem and forces rapid transformation.

In a typical forest, what probably happens is that the preexisting ecosystem is hammered by few years of drought, says Steve Jackson, a paleoclimate researcher from the University of Wyoming. As dying trees topple, the drought literally punches holes in the old ecosystem, giving drought-tolerant species openings to invade. Soon—quite soon by the standards of the "old" ecology—we have an entirely new forest. And rather than the slow, "buffered" changes Behrensmeyer saw in the Omo

---

[56] Michigan isn't the only place to show such changes; 4,000 years ago, Illinois and Iowa had extensive sand dunes. See Paul A. Mayewski, et al, "Holocene climate variability," *Quaternary Research*, 62, 243–255 (2004).

Valley, many of these recent ecosystem shifts are more like the shift in vegetation that occurred in the tropical coal swamps at the end of the Westphalian era—only much, much more rapid.

Such changes aren't limited to prehistory. Jackson knows of a region in New Mexico where a 1950s forest fire killed off a stand of ponderosa pine. Today, trees are growing back, but they aren't ponderosa. They're junipers and pinion pines—representing a major change to a drier, more drought-tolerant ecosystem.

Another example is an area in Central Oregon called the Lost Forest. A few years ago, I backpacked into it, finding a woodland that to all appearances looked healthy—a cloak of ponderosa pine spreading across five square miles of buttes, ashy scablands, and rocky promontories. But the forest is barely hanging on. There are no other ponderosa pine for dozens of miles in any direction and the region draws a scant nine inches of rain per year—significantly less than the amount usually necessary for pine. The Lost Forest is actually a remnant from damper Ice Age climes—a leftover that has managed to perpetuate itself by shading the volcanic soils just enough to allow those nine inches of rain to suffice. But someday, it will be stretched beyond its ability to cope. Insects or fire will kill too many trees. The protective shade will disappear and in the blink of an ecological eye the forest will be truly lost, replaced by the juniper and sagebrush that dominate the surrounding landscape.

When I first researched this article, the American West was in the fourth year of a new drought. The forest fire season, which usually hits my home state of Oregon in late summer and early fall, had begun in earnest in July. In the Rockies, fires had blazed through juniper forests that hadn't burned so severely for at least 800 years. Just as the burned pines in New Mexico are being replaced by junipers, these burned junipers may be replaced by something else.

It's easy to shout "global warming" with each drought or heat wave, but the Earth indeed appears to be in the midst of ecological change more dramatic than anything since the K-T boundary. Some of the causes may be natural—the adjustments one would expect to see on a planet still rebounding from its latest Ice Age—but humans play a major role. Another mass extinction has begun, and it appears to be accelerating.

If there's a core question to be asked of paleoecology, it's this: what happens to ecosystems following such extinctions? What happens to us?

The answers are uncertain. The Permian and K-T extinctions each were followed by explosions of new biodiversity—but not until the Earth had (metaphorically) licked its wounds for many years. One question is what human bioengineers might do to speed the recovery if we find that we've crippled our ecologies and desperately need more diversity. Bringing long-extinct species back to life in the manner of *Jurassic Park* is a theme that's been widely explored in science fiction. But could a more recently extinct species be reintroduced, or would it simply go extinct again, a victim of the same forces that drove it to death in the first place?

The good news from paleoecology is that the Earth has survived some truly severe climate shifts. One we've not previously discussed was spotted in the 1970s in a Massachusetts bog, when researchers observed that 8,200 years ago there was a sudden decline in hemlock pollen and a rise in beech pollen. At the time, the finding was a mystery. But now, scientists know it represented a climate change in which the region suddenly turned colder and drier. And a few thousand years earlier, about 12,000 years ago, there was an even more dramatic climate shift called the Younger Dryas.

The Younger Dryas accompanied the end of the last Ice Age and is named for a cold-climate flower (*Dryas*), whose pollen appeared virtually overnight in European bogs.[57] It was an interlude in which far-northern temperatures dipped by up to 25°F over the course of about a decade, and stayed low for more than 1,000 years, before suddenly rebounding to normal.

The cause of these changes is open to debate (one controversial and probably incorrect theory is that the Earth was hit by a comet or asteroid) but what matters here is their ecological effects, the most important of which is that nothing important seems to have gone extinct. They simply shifted their ranges, very quickly, in response to the climate shift. Spruce trees, for example, declined in Canada's Maritime Provinces but became more abundant in Massachusetts.

Ice core records from Greenland show that in the past few hundred thousand years, the Earth has gone through dozens of similar climate swings, many occurring fast enough to cause severe grief even to technologically sophisticated humans.[58] But for ecosystems, what this means is that unlike those of the Permian extinction and the K-T event, ours today are already survivors of many climate shifts, some much more rapid than anything likely to result from human-caused global warming.

---

[57] There's also an Older Dryas, which occurred several thousand years earlier.

[58] See Richard B. Alley, *The Two-Mile Time Machine: Ice Cores, Abrupt Climate Change, and Our Future*, Princeton University Press, 2000.

When climate shifts, these ecosystems have therefore proven themselves adept at reestablishing themselves in new locations or surviving in refugia comparable to Behrensmeyer's slow-dying canopied forests.

This doesn't mean human-induced climate change won't push many species over the brink. Nor does it mean that ecosystems adapted to the rapid cooling cycles seen in the past will react equally well to today's unprecedented warming. "We live in interesting times," warns Jackson, echoing the views of many other scientists, "and they're going to get a lot more interesting very, very soon." What paleoecology *has* taught us is that we live in a world that has seen its share of interesting times in the past.

FROM FIMBULWINTER TO DANTE'S HELL

The Strange Saga of Snowball Earth

Once upon a time, there was a planet. It was a happy place, as planets go, with sapphire seas and swirling clouds. Then the climate cooled. Snow fell in places it usually didn't and failed to melt in places it usually did. Pack ice formed on once-temperate seas. The snow and ice reflected sunlight back into space and the temperature ratcheted down another notch and then another, as the ice advanced from the poles toward the midlatitudes.

It was by no means the first time this had happened, but always before, the ice had stalled, hesitated, and retreated. This time, it continued to advance, gaining momentum as the planet's albedo—the fraction of incoming sunlight it reflected back into space—rose ever higher. And then, catastrophe. The ice reached a critical latitude—roughly equivalent to Southern California—and the albedo effect escalated until the planet was covered with ice from pole to pole. Average temperatures hovered around -60°F—colder at the poles, not so cold at the equator. Rain ceased. Snow nearly did, too, though flurries still fell, condensing from moisture that sublimated from the relatively warm equatorial ice. If this world was once alive, it was now virtually dead. The once-happy planet had become an icehouse, home only to a few bacteria that managed to eke out a living in the interstices between ice granules. It was so cold that it might, in fact, never re-thaw until its sun grew old and began expanding into a red giant star, hot enough not only to dispel the ice, but soon thereafter to reduce the entire world to molten lava.

This sad little tale stems from a climate model developed in the 1960s by Mikhail Budyko of Leningrad Geophysical Observatory, who observed that if glaciers spread too far toward the tropics, a runaway albedo effect would send the planet permanently into the deep freeze. It was a model that spawned a flurry of space-exploration stories in which starfaring archeologists come upon such planets, finding the remnants of some brave civilization's desperate last stand against the ice. In one story I read in high school, alien explorers eventually discover that the stricken world is our own Earth.

But the model appeared to have no practical ramifications. If, for a science fiction story, you want to create *Fimbulwinter*, the cold hell that in Nordic myth preceded the end of the world, a runaway albedo effect is certainly a great way to do it. But everyone presumed that the real Earth wasn't prone to such catastrophes. Otherwise, one would have already happened and we wouldn't be here to speculate about it.

Then something funny happened. For years, geologists have known that the Earth has seen some large ice ages in its distant past. Outwash deposits from ancient ice sheets have been found in areas as disparate as Africa, Death Valley, and Australia. By the mid-1960s, W. Brian Harland of Cambridge University had identified such deposits on most continents, dating many of them to the Neoproterozoic Era, which ran from about 800 million to about 600 million years ago. Superficially, it looked as though the Neoproterozoic had seen the mother of all ice ages. But for years, most scientists presumed this was an artifact of continental drift. Once the continents' wanderings were fully understood, they figured, it would be discovered that the glacial deposits had been laid down at times when the regions in question had been in the far north or south. After all, everyone knew that the only glaciers that exist in the tropics are small ones on tall peaks such as Mt. Kilimanjaro or the Peruvian Andes. The fact that these deposits seemed to have occurred virtually *everywhere* in the Neoproterozoic was a nagging mystery, but not one that drew much attention.

Then in 1992, Caltech geobiologist Joseph Kirschvink wrote a two-page paper for an obscure scientific monograph, in which he argued that those deposits really meant what they implied. Budyko's catastrophe, he argued, had actually occurred, creating conditions he colorfully dubbed "Snowball Earth." Interest in the idea didn't really take off, though, until Paul Hoffman of the Department of Earth and Planetary Sciences at Harvard

University refined the theory in a 1998 article in *Science*.[59] Since then, the evidence has mounted that Hoffman and Kirschvink were right and that the Earth has gone through a snowball state at least twice: once 750 million years ago, and again 580 million years ago. Hoffman thinks it happened twice again, at intermediate dates—for a total of four events—but the evidence for this is weaker.

The Big Melt

The moment you start taking Snowball Earth seriously, you immediately bang into the main obstacle to Budyko's original model: the albedo effect of planet-girdling glaciers is extremely strong. In his 1992 paper, Kirschvink had to address one crucial question: What melted the ice?

The answer turned out to be a climate factor overlooked by Budyko: carbon dioxide.

Anyone who's paid attention to today's global warming debate knows that carbon dioxide is a greenhouse gas that warms the planet by trapping heat in the air. In the era before fossil-fuel burning, the primary means by which it entered the atmosphere was from volcanic eruptions. Excess was taken out of the air by dissolving in rainfall to form carbonic acid, which

---

[59] Hoffman's original article is "A Neoproterozoic Snowball Earth," *Science*, 28 August 1998, pp. 1342-46. For non-geochemists, it's tough going. For more accessible summaries, see:

• "Snowball Earth," Paul E. Hoffman and Daniel P. Schrag, *Scientific American*, January 2000, pp. 68-75.

• "An Appealing Snowball Earth That's Still Hard to Swallow," Richard A. Kerr, *Science*, 10 March 2000, pp. 1734-36;

• "The Longest Winter," Gabrielle Walker, *Natural History*, April 2003, pp. 44-51; and

• "Ice Magic," Gabrielle Walker, *New Scientist*, 12 April 2003, pp. 30-34.

reacts with certain types of rocks, specifically silicates and carbonates, forming byproducts that wash into the ocean.

During the Snowball, volcanoes continued to erupt, melting holes in the ice, through which carbon dioxide spewed into the atmosphere. But the Snowball would have largely shut down the carbonate and silicate weathering cycle by burying the rocks in ice and snow. Even if there were gaps, the weathering process is temperature dependent, proceeding much more slowly under Snowball Earth conditions. Not to mention that there would be no rain to bathe weatherable rocks in carbonic acid.

Today's volcanoes produce about half a billion tons of carbon dioxide per year. If Neoproterozoic volcanism proceeded at the same rate, Hoffman estimated that it would take about four million years to accumulate enough carbon dioxide in the air to reverse the Snowball—somewhat longer if the weathering cycle wasn't completely shut off or if there were cracks in the sea ice that allowed some of the carbon dioxide to dissolve directly from the air into the ocean. In his original article, Hoffman estimated that the build-up might take somewhere between nine million and 30 million years.

However long it takes, the critical value for the amount of carbon dioxide needed to reverse the Snowball was about 120,000 ppm, he calculated. That's a whopping 12 percent—a level hundreds of times higher than those that have us concerned today. What this means is that once the Snowball started to melt, it must have melted fast—so quickly that *all* of the ice may have been gone within a few hundred years. Meanwhile, all of that carbon dioxide was still hanging around, with no gigantic albedo effect to offset its warming potential. In the blink of a geological eye, the Earth didn't just rebound from the Snowball: it overshot, increasing its average temperature by approximately 180 degrees, from -60°F to 120°F. Goodbye icehouse, hello steamhouse. Dante, Hoffman says, would have been pleased.

Death of a Supercontinent

The weak link in the Snowball theory is the question of what might have started the runaway glaciation. Earth has seen many ice ages, but since the Neoproterozoic there is no evidence that one ever again got out of hand. What was different back then?

One factor is that the Sun was weaker. Aging stars put out more energy than young ones, and the 750 billion years since the Neoproterozoic is enough for the Sun's output to have intensified by six percent. But earlier in the Earth's past, the sunlight was weaker yet, and while there is evidence for another Snowball era 2.5 billion years ago,[60] such episodes have not been the norm. Other factors must have been at work.

In a 2002 paper in *Geochemistry, Geophysics, Geosystems,*[61] Hoffman and his colleague Daniel Schrag argued that one of these was the arrangement of landmasses. During the Neoproterozoic, most of the Earth's land appears to have been arrayed near the Equator, where it would have had a quite different effect on climate than today's arrangement of continents. To begin with, land has a higher albedo than water. Under the intense sun of the tropics, that albedo will have a greater effect on climate than when the land is farther north or south, making the Earth more susceptible to cooling. In

---

[60] D.A. Evans, et al, *Nature,* 1997, pp. 262-265.

[61] Schrag, Hoffman, et al, *Geochem. Geophys. Geosyst,* 3(6), pp. 1-21 (June 2002).

addition, all of that equatorial land may have played hob with ocean currents, which today convey a considerable amount of heat from the tropics to the poles. Also, when glaciations begin on today's Earth, with its high-latitude landmasses (such as Canada, Alaska, and Siberia), there is an immediate reduction of the carbonate/silicate weathering cycle as the ice covers weatherable rock. *I.e.*, the moment these areas start becoming glaciated, carbon dioxide begins building up in the atmosphere. In the Neoproterozoic, however, weathering would have continued unabated, because the polar ice would have been mostly sea ice. Once the ice age had advanced far enough for ice to begin forming on land, it would have been too late for small changes in the carbon-dioxide weathering cycle to halt its advance.

An even stronger argument was proposed two years later by a team led by French climate researcher Yannick Donnadieu, who noted that the first Snowball glaciation actually coincided with the *breakup* of a supercontinent called Rodinia.[62] Generally, the interiors of continents are dry, like the steppes of Asia or America's western Great Plains. In a supercontinent, the interior may have been very dry indeed.

But when Rodinia started breaking apart, land that had once been in the interior became coastal, exposing a lot of previously unweathered rock to more rainfall. Donnadieu estimated that

---

[62] Donnadieu, et al, *Nature* 18 March 2004, pp. 303-306. Another factor may have been that the breakup of Rodinia was accompanied by huge flows of *flood basalts*—a type of lava that spreads across an immense area. The eruptions producing these basalts would have caused a short-term spike in atmospheric carbon dioxide, but the huge amount of fresh, easily weatherable rock would soon have drawn more carbon dioxide out of the air than the eruptions injected into it.

this was sufficient to cause the Earth's atmospheric concentration of carbon dioxide to plummet from pre-breakup levels estimated by geologists to be 1830 ppm to less than 250 ppm over the course of a few million years. That would've been enough, he calculated, to produce a global cooling of about 8°C (14°F)—and far more than enough to launch one heck of an ice age.

Where's the Evidence?

Climate models are a dime a dozen. What makes the Snowball Earth idea interesting is that there's considerable evidence it's correct. Unfortunately, some of that evidence is the type of thing that makes my English-teacher mother glad nobody ever forced her to take a course in geophysics: a nasty-sounding stew of terms like *paleomagnetism, cap carbonates,* and $\partial^{13}C$ *values*.

The story starts simply enough, though, with a visit to one of the world's most exotic landscapes, Namibia's Skeleton Coast. That region, rapidly developing into a tourist destination, is close to southern Africa's sprawling Kalahari Desert—though the arid lands of Namibia are called the Namib Desert. Geologists love arid country because (as someone once told me) the land isn't all "haired over" with plants, making it easy to see the underlying geology. Add in some dramatic rock formations, and you have geologists drooling, right along with the ecotourists.

There are lots of things to study on the Skeleton Coast, but the ones that have caught the Snowball Earth theorists' eyes are Neoproterozoic glacial deposits. We noted earlier that such deposits occur worldwide, but they are particularly well studied in Namibia, which Hoffman has been traipsing around since

172

1993.

Seven hundred fifty million years ago, that part of Namibia was submerged beneath a shallow tropical sea, which means it was accumulating sediments throughout the critical Snowball Earth period. One layer is a gray siltstone: a type of stone formed from muddy seabed. Siltstones aren't unusual, but this one is peppered with rocks of all sizes, ranging from pebbles to boulders, and varying widely in composition. The question is: how did all of these rocks get into a muddy seabed? Hoffman's answer: they floated out to sea in icebergs that calved off of a disintegrating glacier. As the icebergs melted, the rocks, called dropstones, fell out of them and sank. These, in fact, are the types of features that led Harland to start cataloging glacial deposits all those years ago. (An alternative theory is that these jumbled deposits were created by submarine landslides, but there are subtle differences between landslides and ice-rafted debris, and even Snowball Earth's strongest skeptics generally accept that these deposits are indeed glacial.)

The next question is determining the latitude at which they were created—something geologists refer to as their paleolatitude (*paleo-* meaning "old"). In principle, this is easy. In practice, it isn't. The technique hinges on the fact that when siltstone forms, magnetic particles in the mud line up with the Earth's magnetic field like tiny compass needles, leaving a magnetic imprint on the resulting rocks. Near the poles, the magnetic field has a strong up-and-down component; near the Equator, it runs more closely parallel to the Earth's surface. The angle at which a rock is magnetized can therefore reveal the latitude at which it was formed.

That's the theory. The reality is that the passage of time is extremely good at scrambling these magnetic imprints. Whenever geological processes melt or partially melt a rock, it becomes remagnetized, now carrying the magnetic signal of the latitude at which it resolidified, not the one at which it originally formed. In addition, water percolating through porous rock can bring in new magnetic particles whose orientation reveals the latitude at the time *they* were deposited.

In 750 million years, a lot of rocks get remagnetized. In 2000, for example, Dave Evans of Yale University reviewed the data from 85 Neoproterozoic glacial deposits. Only 16, Evans found, carried enough residual magnetism to allow *any* meaningful guess as to the latitude at which they were formed, and of these, nearly half provided paleomagnetic data that was at best "poor."[63] But—and this is a huge "but"—among the deposits for which the paleomagnetic data was strongest were ones that appear to have been formed very close to the Equator. Included in that list are the deposits in Namibia, but even more impressive are ones in Australia, originally studied by Nicholas Christie-Blick, a sedimentologist at Columbia University's Lamont-Doherty Earth Observatory.

In 1999, Christie-Blick and colleagues reported that samples from the Australian deposit show not only low-latitude magnetization but at least three polarity shifts in the Earth's magnetic field (which occur at intervals on the order of tens to hundreds of thousands of years).[64] If these sediments had at some time been remagnetized, the entire deposit should have been remagnetized, wiping out any trace of these magnetic flip-

---

[63] D.A.D. Evans, *American Journal of Science* 300 (5): 347–433 (2000).

[64] Linda E. Sohl, Nicholas Christie-Blick, & Dennis V. Kent, *Geological Society of America Bulletin*, 111(8), pp 1120-1139 (1999).

flops, and the fact that signs of these flip-flops remain is strong evidence that they retain their initial magnetism. It appears that during the Neoproterozoic, there were indeed massive glaciers in the tropics.

Such glaciers indicate a massive ice age, but not necessarily a planet-girdling one. Nick Eyles of the University of Toronto argues that to the extent we really are seeing glacial deposits, they're from regional glaciations caused by tectonic uplifts that created tall, glaciated mountains as the Rodinia supercontinent "unzipped."[65] One difference between this and the Snowball is that a limited glaciation of this sort wouldn't shut down the silicate and carbonate weathering cycle, which means there wouldn't be a severe hothouse overshoot following the defeat of the glaciers.

That's a critical difference because the Snowball model, with its subsequent hothouse warming, predicts that as the world suddenly heats up, increasing rainfall would bring huge amounts of carbonate weathering byproducts into the seas, where they would react with other chemicals and precipitate out as massive layers of carbonate rocks such as limestone. And on first blush, that's exactly what we see. Above the gray siltstones of Namibia and other Neoproterozoic glacial deposits lie thick layers of limestone, now referred to as cap carbonates. In his 1998 *Science* article, Hoffman calculated that in Namibia, these cap carbonates must have accumulated at the staggering rate of 40 centimeters per year, producing beds that are now tens to hundreds of meters thick. Overall, he calculated, there was enough excess carbon dioxide in the air to blanket the entire

---

[65] N. Eyles & N. Januszczak, *Earth-Science Reviews* 65 (1–2): 1–73 (2004).

planet in five meters of limestone, though most of it would have formed in shallow-water, coastal regions such as ancient Namibia.

More geochemical evidence comes from iron deposits laid down during the Snowball epoch.

Iron enters the ocean from undersea hot springs. Early in the Earth's history, before there was oxygen in the air, oceanic iron concentrations were quite high. But then photosynthesizing plants began creating oxygen, which reacted with the dissolved iron to form insoluble iron oxides. These precipitated out to form deposits called redbeds, for which modern humans should be immensely grateful because they provide our best iron ore. But all of that oxygen also meant dissolved iron no longer built up to high levels. Subsequent sediments still contain iron, but generally it is dispersed throughout them at low concentrations.

The exception to that "generally" was during a brief interval of the Neoproterozoic, when the redbeds mysteriously reappear. For years, geologists were mystified, but the Snowball Earth theory provides a simple answer because during that time, sea ice would have cut off the oceans from the air—and from sunlight for oxygen-producing plants. Oxygen remained in the atmosphere, but in the water its concentration dropped. Iron, on the other hand, continued to percolate up from submarine hot springs and its concentration began to mount. Then the Snowball ended. Oxygen returned to the ocean and the iron abruptly precipitated out. Almost by accident, the Snowball theory explains yet another geological mystery.

The final question is how all of this affected life. Initially, Hoffman argued that life suffered dramatically—just as one might expect. The evidence was based not only on the apparent

176

deoxygenation of the ocean, but on the ratios of carbon-12 to carbon-13 in carbon-containing sediments—carbonates—laid down just before the Snowball sealed the oceans.

Carbon-12 and carbon-13 are stable isotopes (unlike carbon's radioactive isotope, carbon-14, used for carbon dating). The Earth contains them in a fixed ratio, but biological reactions tend to work slightly better with the lighter isotope, carbon-12. The result is a fractionation process in which biological processes such as photosynthesis produce products that are slightly carbon-13 depleted, while nonbiological ones do not…meaning that the relative amount of carbon-12 and carbon-13 in any carbon-containing substance ($\partial^{13}C$) is an indicator of the role life played in its formation.

The scientific literature on this subject is complex, but the basic point is that Hoffman argued that carbonates laid down just before the height of the Snowball were unusually low in carbon-12, indicating that life was already having a hard go of it. But it's not quite that simple. A group led by Christie-Blick—never, as we will see, a true believer in the Snowball—claimed that they knew of carbonate sediments laid down during the putative Snowball era whose $\partial^{13}C$ ratios indicate that life was doing just fine.[66] Hoffman and Schrag countered that the acidic seas created by the Snowball would dissolve *old* carbonates, loading the ocean with their organic-based carbon. This, they said, later mixed into some of the later carbonates, producing a false impression that they were organically derived.[67]

---

[66] Kennedy, M.J., Christie-Blick, N. & Prave, A.R., *Geology* 29, 1135-1138 (2001).

[67] Halverson., Hoffman, Schrag, et al, *Geological Society of America Bulletin*, 117, 1181-1207 (2005).

HERE BE THERE DRAGONS

By 2004, the debate had crystallized into competing snowball visions that persist to this day. (The best synopsis of the competing papers is on Hoffman and Schrag's website, www.snowballearth.org, though there hasn't been a lot of recent activity; the debate peaked in the early 2000s.)

One theory is the "hard" Snowball model, in which the Earth has no open water whatsoever. The other is a "slushball," in which the ice didn't quite reach the Equator and the world never quite froze over completely.

The Slushball model has some important things to recommend it. To begin with, it solves a major problem with some of the Snowball deposits. In a hard Snowball, the Neoproterozoic glacial deposits should have been laid down quickly during the big melt…and there should have been only one such melt. But that's not what the data show. Christie-Blick's Australian deposits were laid down over a long-enough interval to preserve several magnetic flip-flops. And a 1,500-foot-thick deposit in Oman appears to have been created in multiple cycles of glacial advance and retreat.[68] Slushball Earth provides a mechanism by which this could happen.

Furthermore, it's possible that Slushball elements such as thin ice at the Equator and open-water leads might have allowed sunlight to reach photosynthetic bacteria. In 2011, Adam Campbell of the University of Washington, Seattle, and colleagues argued that all that's needed is a long, narrow ice-free

---

[68] *See* "Snowball Melting?" C. Marzuola, *Science News*, October 19, 2002, p, 246, and more recently, Rieu R, Allen PA, Ploetze M, et al. (2007), "Climatic cycles during a Neoproterozoic 'snowball' glacial epoch," *Geology*, Vol. 35, 299-302.

body of water, comparable to today's Red Sea.[69] Other refugia might have existed at surface-level hot springs and undersea hydrothermal vents, or there may have been open-water rifts where sea ice was deflected away from the edges of continents or islands.

Ediacarans, the Snowball, and Us

The Snowball Earth theory also opens the door to the idea that the Snowball—or more precisely, the final one (580 million years ago)—is what created life as we know it.

Before that time, life wasn't really all that interesting. All that existed were bacteria and algae, which lived in colonies that Gabrielle Walker, author of *Snowball Earth: The Story of the Great Global Catastrophe that Spawned Life as We Know It* (Bloomsbury Publishing PLC, 2003) colorfully described as "Slimeworld." Then, something changed. Single cells learned to cooperate, and Slimeworld was replaced by the Cambrian explosion, in which life suddenly became *life*, with all its wondrous complexity.

One line of evidence linking the first development of multicellular animals to the Snowball epoch comes from "molecular clock" studies in which biologists estimate the rate at which related species diverge. In 2002, Kevin Peterson of Dartmouth College compared proteins from sea worms, sponges, and echinoderms (the animal family that contains sea stars and sea urchins), discovering that they—and apparently all multicellular animals—radiated from a common ancestor that lived about 700 million years ago—nicely in the middle of the

---

[69] Campbell, et al, *Gephysical Research Letters,* Vol 39, L19502 (2011).

Snowball period.[70] The first fossil evidence for multicellular animals, however, comes from slightly later, during a pre-Cambrian period called the Ediacaran.

The Ediacaran has the distinction of being the first newly recognized geological period since the 1880s. The International Union of Geological Sciences formalized its designation in 2004, after a bitter fight that revolved not around whether the period was distinctive enough to be recognized as a separate subunit (there was little dispute about that) but whether it should be given the name Ediacaran (favored by Australians) or Vendian (favored by Russians). Geology has long been a science in which egos reign strongly.

The Ediacaran began 600 million years ago and ended at the start of the Cambrian Period, 545 million years ago. This means it has the distinction of including the final Snowball epoch.

One of the leading experts on the Ediacaran is Guy Narbonne, from Queen's University in Kingston, Ontario, whose research has often focused on rocks found in southeastern Newfoundland, buried under ash from an ancient volcanic eruption.

To geologists, volcanic ash is one of nature's greatest creations. Normally, fossils are rare and scattered, at best providing snapshots of ancient ecosystems. But ash buries everything all at once. Just as archaeologists have been able to learn much about early Rome by digging up the ash-buried ruins of Pompeii, fossil-containing ash beds are treasure troves of ecological information for geologists. The Newfoundland beds

---

[70] Andrew Smith & Kevin Peterson, *Annual Reviews of Earth and Planetary Science*, 2002, pp. 65-89.

reveal seabed creatures looking like nothing known today.[71] Some were flat fronds. Others were comprised of long stems, anchored to rocks. Still others grew as cabbage-like bushes. The vast majority were constructed of fractal blocks that make them look like they'd been designed by a child playing with Legos. Narbonne believes that these were bottom-dwelling animals, not plants: filter feeders that held fast to the rocks and grazed on passing algae and bacteria. His argument is based on their height distribution. Plants normally compete to rise high in the quest for sunlight. Filter-feeding animals find most of their food near the bottom, and are relatively short—exactly the case for Narbonne's Ediacaran community.

Evolutionary biologists have advanced several theories for why the Snowball might have triggered a massive evolutionary leap from single-celled to multi-celled life. The simplest is that it wiped out most prior life, forcing the survivors into isolated pockets that allowed them to randomly diverge until those in one of these pockets stumbled onto multicellularity. It's a theory that works best with small, very isolated populations capable of rapid genetic change, making it more appropriate for a hard Snowball than a Slushball, where the isolation may not have been complete.

Another theory is that after the Snowball/Slushball melted and the hothouse had surged and abated, the Earth had a lot of vacant ecological niches into which any organisms that happened to be around could expand and flourish. Briefly, competition abated and new genetic experiments had a chance to take root. An alternative version of this theory says that the

---

[71] Narbonne & Gehling, *Geology*, 31(1), pp. 27-30 (January 2003).

Snowball/hothouse yo-yo was an evolutionary sieve through which multicellular organisms could've passed more easily than their predecessors. Such organisms may have first appeared millions of years earlier, as Peterson's molecular clock indicates, but it wasn't until after the wild changes during and after the Snowball that they could come to the fore.

A third theory says that their growth was facilitated by an upsurge in oxygen following the post-Snowball Hothouse. During the Hothouse, the oceans would have been a stew of nutrients promoting riotous algal growth that could easily have returned us to Walker's Slimeworld. But all of that algae could also have released a tremendous amount of oxygen into the air. This would work to the advantage of multicellular organisms, which need a lot of oxygen. If there's not enough, they're at a disadvantage, but when it's plentiful they can use it to be stronger, faster, and bigger than their single-celled competitors. In a world with lots of oxygen, multi-cellular organisms are like Olympic athletes. Single-celled bacteria and algae still exist—but they're still slime.

Did the Snowball really happen? It's one of those theories that, from a science-fictional perspective, is so appealing it's hard to judge objectively. But it's clear that something strange happened, and that this something involved a lot of ice.

The Snowball also raises a question that has nothing to do with geology. One of the enduring debates in science fiction is whether complex life exists on other planets. If the geological weirdness of the Snowball was the cue that jolted our own planet's life out of single-celled complacency, complex life elsewhere may be scarcer than we think. Not only do you need a suitable planet, you also need one which, well after life has taken hold, goes through comparably wild fluctuations…then stabilizes again so that continuing oscillations don't keep killing off whatever interesting lifeforms might have developed.

The fun in science fiction, of course, is trying to come up with new and innovative ways to do this. Earth used the Snowball. Are there other processes that might achieve something similar?

WHY EARTH ISN'T MARS OR VENUS

Plate Tectonics, Goldilocks & the Late Heavy Bombardment

In my part of the world, weather forecasts are the subject of much joking—hard to get right, sometimes even as little as 24 hours in advance. But in the billion-year time frame, the Earth's weather is easier to predict. Warm but not too hot, cool but not too cold. Not much chance of freezing the oceans solid or evaporating them away completely.

This didn't have to be the case. Mars and Venus appear to have started out quite earthlike but then veered into climate dead-ends: one frozen, the other baking in a runaway greenhouse. Why then is Earth's climate so stable? Even when something tips it enormously out of whack—like the giant ice age of Snowball Earth (see Chapter 9), our planet recovers.

Partly, of course, this is a side effect of a lucky position in the Solar System. Not only are we in the Sun's habitable zone, where planets won't automatically wind up like Mercury or Pluto, but of the three worlds in this zone, we're the Goldilocks planet.

"In very simple terms," says Victor Baker, a planetary geologist at the University of Arizona's Lunar and Planetary Laboratory, "Venus is too hot, Mars is too cold, and Earth is just right." But that's just the starting point. In the 4.5 billion years since the Solar System formed, the Sun has brightened by about 25 percent, but our climate has barely changed. Obviously, something is keeping it stable.

In the old days, scientists and science fiction writers credited the Earth's oversized moon with our planet's continued habitability. The theory was a bit vague, but was based on the observation that the Earth and Moon are a unique pairing. Venus is moonless, and the two moons of Mars, Phobos and Deimos, are tiny. Maybe, the theory went, life-supporting planets, here or in other solar systems, might need big moons. And in fact, the latest theories indeed suggest that the moon plays a role—something we'll discuss later. But two other factors are more important: plate tectonics and water. In combination, they interact with carbon dioxide to form a very effective global thermostat.

This process begins with volcanoes. Volcanoes, as we discussed in the previous chapter, spew out more than molten rock. They also produce gas, a major component of which is carbon dioxide. As anyone familiar with the global warming debate knows, carbon dioxide is a "greenhouse gas" that traps infrared radiation that would otherwise escape to space. Glass does the same thing, which is why greenhouses can stay warm

186

without heaters and why the inside of your car gets so hot in the summer. Without greenhouse gases such as carbon dioxide, scientists calculate that our planet would not have been able to rise above the freezing point for much of its early history. But for a habitable planet, they can't just build up forever, so something needs to remove excesses, especially carbon dioxide, from the atmosphere.

That something is rain, which (again as we discussed in the previous chapter) dissolves carbon dioxide from the air, creating a weak acid that falls on exposed rocks, reacting with them in a process called atmosphere weathering. This not only slowly dissolves the rocks, but creates carbon-containing byproducts that wash out to sea, where they eventually accumulate on the seabed as carbonate rocks, such as limestone. This gives us two opposing processes. One (volcanoes) injects carbon dioxide into the air. The other (rain) removes it. Furthermore, the two processes react differently to changes in temperature. Volcanoes are unaffected by it; over the long run, they inject carbon dioxide into the atmosphere at a more-or-less steady rate. Atmospheric weathering, on the other hand, is highly dependent on temperature. When the planet is hot, more water evaporates from the oceans, which means that more falls as rain, increasing the rate of weathering. When the planet is cooler, the process slows down, allowing volcanic carbon dioxide to build back up. The result: a thermostat that allows for even big climate fluctuations to revert to a happy medium.[72]

---

[72] Ideally, we could rely on this process to abate our current climate concerns. Unfortunately, it's slow, requiring thousands of years to adjust.

Recycling the Seabed

So far, though, we've not described a process that can operate on a billion-year timeframe. Without a way to replenish the carbon dioxide going into volcanoes, they'd eventually run out. That's where plate tectonics enter the picture.

In simplest terms, plate tectonics is the cause of "continental drift." That's because the Earth's crust is composed of plates that slowly move in response to deep currents in the Earth's mantle. Where these currents rise to the surface, plates get pulled apart. Elsewhere, plates collide. The part of this process that's relevant to the global thermostat is "subduction," in which seabed plates are shoved beneath continental plates, pushing the seabed into the Earth's interior, carrying carbonate sediments with it. There, the Earth's heat bakes the carbon dioxide out of them, allowing it to mix with magma, rise to the Earth's surface, and erupt in volcanoes.

We now have a simple but important cycle: volcanoes, carbon dioxide, rain, atmospheric weathering, carbonate rocks, subduction, and more volcanoes. It's a cycle in which carbon dioxide is not only used over and over again, but the Earth's temperature tends to stay near that happy medium. And, it's a process that's operated on Earth through most of our planet's history and is likely to continue into the distant future—so long as we continue to have both plate tectonics and a liquid ocean. But it's not operating on Mars and Venus. Why were we lucky, when our neighboring planets weren't?

Disappearing Water

In the case of Venus the answer probably lies in the fact that there's a limit to how much solar heat the water/tectonics/carbon-dioxide cycle can accommodate. Although there are many mysteries about Venus, scientists think it probably started out very much like the Earth, with liquid water, rainfall, and plate tectonics. But it was just enough closer to the Sun to get in trouble.

The problem was that carbon dioxide isn't the only greenhouse gas. Water vapor is, too. Thus, there are two competing effects when an ocean-bearing planet gets warmer. One is the carbon dioxide/rainfall thermostat. But opposing that is the effect of increasing water vapor. At low-to-moderate temperatures, that's minor. But as the planet warms, the amount of water vapor in the atmosphere rises exponentially. "That makes for a stronger greenhouse, which leads to a hotter surface and more evaporation,"
says David Grinspoon, curator of astrobiology at the Denver Museum of Nature & Science. "Once that runaway's going, it's hard to stop."

Eventually, Venus became so hot that essentially all of its water was in the atmosphere as vapor, producing a super-greenhouse that kept the surface too hot for rain to fall. That, in turn, allowed carbon dioxide to build up, making the planet hotter yet. Meanwhile, water vapor was making its way into Venus's upper atmosphere, where solar radiation could break it into hydrogen and oxygen (a process that has been observed going on today by instruments on the Venus Express spacecraft, currently in orbit around Venus). Hydrogen is too light a gas for terrestrial-sized planets to retain, so it escaped into space, leaving Venus with only traces of its original water. "That means no more thermostat," says Grinspoon. "You don't have any way to cool off."

Mars has a different story. Scientists once thought that it, like Venus, had lost its water early on—or that it had never had much to begin with. But from the moment we started flying probes to the Red Planet, that theory started looking doubtful. To start with, Mars has thousands (and probably millions) of miles of river channels. In a 2008 study (updated in 2010)[73], a team led by Brian M. Hynek of the University of Colorado's Laboratory for Atmospheric and Space Physics painstakingly counted all the channels revealed by the latest high-resolution satellite photos. Their tally: 40,005 distinct river valleys. Not only did this quadruple the known number of valleys, but some were seventh-order streams, meaning they were tributaries of tributaries of tributaries . . . all the way up though seven tiers of stream branchings. That's important because it indicates that these watersheds were formed by a dispersed source, rather than scattered springs. In other words, by rain.

But simply counting valleys wasn't Hynek's main goal. Rather, he asked, "Can we figure out the last time it rained and formed valleys on Mars?" The answer is a qualified yes. The new satellite images are good enough that it's possible to count the numbers of craters in creek beds—a standard technique for determining the age of planetary landscapes. The more craters, the longer an area has been exposed to meteor bombardment, and the older its surface must be. Hynek concluded that some creek beds may have been formed as recently as half a billion years ago, but these look to be of the type created by seepage from springs. The last ones that were clearly precipitation-fed

---

[73] Hynek et al, *J. Geophysical Research: Planets*, 22 Sept 2010.

were formed about 2.8 billion years ago. That's a long time ago, but Mars is 4.5 billion years old. Thus, for more than one-third of its history, it had enough water for at least scattered rainfall.

Nor has this water vanished. We've long known that Mars has ice caps, but they're only a fraction of the amount of water needed for an earthlike planet. More water appears to be hidden in permafrost, some quite close to the surface. In 2008, for example, the Phoenix lander found ice crystals in a trench scratched only a few inches into the surface of the Martian arctic. Additional permafrost is widely believed to lie more deeply buried, elsewhere. There may even be mountain glaciers buried under thin layers of rock and dust. In several places in the Martian mid-latitudes, roughly between 35° and 55°, either north or south, oddly lobed "debris aprons" spill from crater rims. Photos show them to be rock covered, but their shapes suggest they were formed by something flowing downhill, something that behaved a lot like glacial ice.

A few years ago, scientists got a chance to look at two of these debris aprons with ground-penetrating radar carried on NASA's Mars Reconnaissance Orbiter. In a 2009 paper in *Geophysical Research Letters*,[74] Jeffrey Plaut of NASA's Jet Propulsion Laboratory and an international team of coworkers found that whatever lies beneath the rocky surface is virtually transparent to radar to a depth of about 500 meters, indicating it's probably made of 90 percent pure water ice. The volume of ice wasn't huge, but it was close to the surface (within 10 meters, Plaut estimated from the radar signal): yet another sign that a great deal of water still exists on Mars, scattered around the

---

[74] Plaut, et al, *Geophysical Research Letters*, 36(2), January 2009.

planet in icy deposits. All of this, plus the copious river channels, suggests that Mars spent its first half-billion years wet and fairly warm. "It may well be that Mars, early in its history, had something like [the Earth's] thermostat," Grinspoon says.

And whatever killed the thermostat wasn't a loss of water. Rather, it appears to have been a freezing-up of plate tectonics.

One of the effects of plate tectonics on Earth is the existence of paleomagnetic "stripes" in the seabeds, marking bands of volcanic rock that oozed up along mid-ocean ridges as tectonic forces gradually widened the seabed. Each stripe carries the imprint of the Earth's magnetic field at the time its rocks solidified—and because the magnetic field varies with time (see Chapter 9), the magnetic pattern also varies, with similar-age rocks having the same imprint. Using this, geophysicists can trace the history of the ocean's spreading.

By 1999, scientists were finding similar magnetic striping in the oldest rocks of Mars: a strong sign that these rocks had been formed in a similar manner, as magma oozed up from below to fill gaps formed by the spreading of tectonic plates.[75] But if Mars was indeed born with plate tectonics, they appear to have shut down quite early in its history. There are certainly no signs of geologically recent tectonic activity.

To understand what went wrong, we need to look deeper into the planet's interior.

Plate tectonics as we noted earlier, is driven by currents in the mantle. But these are themselves driven by heat escaping from the planet's core, so ultimately it is heat from the core that is responsible for tectonics.

---

[75] See Richard A. Kerr, "Signs of Plate Tectonics on an Infant Mars," *Science* 30 April 1999, pp. 719-722.

Mars is substantially smaller than the Earth, which means its core is also smaller. When it was young, that wouldn't have made much difference, but with the passage of time, its core would have cooled more quickly than the Earth's, reducing the heat supply to the mantle. Less heat flow means less-vigorous mantle currents, which in turn means less-vigorous tectonics and a crust that steadily thickens as it, too, cools. Eventually, the crust becomes too rigid to move, and tectonics grind to a halt. Grinspoon compares Earth and Mars to baked potatoes. "Big ones stay hot longer," he says. "It's the same with planets."

Death of A Dynamo

The difference in size also made it hard for Mars to retain a thick atmosphere. "It probably lost more than 99 percent of its original atmosphere," Grinspoon says. "You simply can't support a strong greenhouse effect when you have that thin an atmosphere."

Traditionally, that loss was blamed on gas escaping the planet's weak gravity. But the core may also have played a role by switching off the Martian magnetic field early in the planet's history.

Planets' magnetic fields are thought to be created by convection currents in molten-iron cores—currents that cause electrically charged particles trapped within them to act as a dynamo. This, in turn, induces a magnetic field in a process related to those used in everything from electrical power generation to electric motors.

We know that Mars once had a magnetic field thanks to the magnetic stripes seen in the early rocks. But later rocks aren't magnetized, and by comparing them, we can estimate that the dynamo shut off when the planet was only a few hundred million years old. This is important because a planet's magnetic field does more than simply imprint its signature on solidifying lava. It also shields the planet from charged particles in the solar wind—ones, which, unblocked, can speed the dissociation of water in the planet's atmosphere.

Mars isn't Venus, so we know it never got hot enough to vaporize more than a small fraction of its water. But the water that was in the air when the dynamo stopped was the part that was contributing to the planet's greenhouse warming. Thus, as its core cooled, Mars was hit by a double whammy. At ground level, plate tectonics shut down, reducing the supply of fresh carbon dioxide. Higher up, water vapor was being lost to space. Between them, these two processes cooled the planet enough that eventually its crust thickened too much for tectonics to continue, giving us the tectonically inactive planet we know today.

An alternative theory is that the dynamo may have been struck a fatal blow during the late heavy bombardment, a phase near the end of the planet-forming epoch when the last major collisions were truly *big*. On the Earth, one of these, with a Mars-sized object, may have formed the Moon. On Mars, a collision with a 1,000-mile to 1,800-mile object may have created the Borealis Basin: a northern lowlands covering roughly 40 percent of the planet. Overall, says James Roberts of Johns Hopkins University Applied Physics Laboratory in Laurel, Maryland, we can count about 20 really large impact basins on the Martian surface, assigning dates to them by comparing the numbers of smaller impact craters on their floors. (Not surprisingly, a lot of the large craters turn out to be 4.1 to 4.2 billion years old, representing the heaviest part of the late heavy bombardment.) But we can also use satellite measures to determine the degree to which these basin floors became magnetized when they resolidified after the impacts that formed them. And when that's done, Roberts reported at the 2014 Mars conference in Pasadena, California, a striking pattern emerges. The oldest craters have magnetized floors; the later ones do not. Sometime during the late heavy bombardment, the Martian magnetic field seems to have shut off—rather abruptly, it would appear.

Is that a coincidence, or did asteroid impacts kill the dynamo?

Possibly it did, says Roberts. The mechanism would have been the heat of impact. By warming the crust and the mantle, Roberts says, this would have greatly reduced the temperature gradient between the core and the surface, possibly even reversing the direction of heat flow temporarily, with heat going downward rather than upward. And according to his computer models, a weak dynamo turns out to be easy to shut down forever. "You can kill it with a very small drop in heat flow," he says, "but it takes a lot to restart it." The same aberration in heat flow might also have affected mantle currents for long enough that they, too, did not recover. Thus, a single giant impact in the late heavy bombardment (or a succession of impacts) may have been all it took to shut down the Martian magnetic field, setting in motion the series of events that eroded water from the Martian atmosphere, helping to kill its carbon-dioxide-producing tectonics, and sending the planet into its frigid present.

Wayward Spins

So far, we've found that the Earth was fortunate in two ways:

it wasn't too close to the Sun, and it was large enough for plate tectonics and a core dynamo to continue for billions of years. Three other factors also helped.

One is simply the fact that the Earth has life. Life has radically changed the Earth. "If you had done an environmental impact statement four billion years ago and said, 'Should we let life start?' you would have said, 'No,' because it's going to completely screw up the environment," jokes James Head III, a planetary scientist from Brown University. "[Life] has radically changed [the environment]—for the better, we might say, but who knows what was asking the question then!" One of the things life has done is infuse the atmosphere with oxygen, something that wouldn't have happened without photosynthetic plants and bacteria. But that's not the only thing. "A lot of [marine] organisms secrete calcium carbonate shells and they get deposited in the ocean," says Head. "That takes a huge amount of carbon dioxide out of the atmosphere."

The second factor that has helped keep Earth temperate is simply the continued existence of large quantities of liquid water. Partly that water helps because the oceans are a huge heat sink, capable of damping short-term variations in climate. But some of their water is also dragged beneath the surface in subduction zones via waterlogged sediments, where it helps lubricate the process of plate tectonics. "It's not well known how it acts," says Michael Mischna, a planetary scientist at NASA's Jet Propulsion Laboratory, "[but] it allows plates to slip and slide against each other. When you lose that, it locks everything up, like an engine seizing."

The third factor is the Moon. The old theory was that its gravity somehow stripped off excess atmosphere, keeping us from getting Venus-hot. The new theory says that the Moon helps keep ice ages from becoming too severe.

The best way to see how this works is to look at Mars. Mars, Head and other scientists are finding, has long undergone substantial wobbles in its spin axis—wobbles that have wreaked havoc on its seasons. Modeling the planet's motions backward in time, he says, it's possible to calculate the effect of perturbations from other Solar System bodies on the tilt of its axis, finding that in the past few million years it's ranged from very little tilt to an angle of 35 to 40 degrees. In fact, he says, there's a reasonable chance Mars could practically flip over on its side, tilting by 70 or 80 degrees.

The Earth also sees variations in its tilt, but only by a few degrees. The difference, Head and others say, is that the Earth's spin is stabilized by the Moon. "Mars has just these two little space potatoes, Phobos and Deimos, which don't really do much," Head says. Mischna adds: "Without something like the Moon, essentially locking the Earth into place, the axial tilt tends to wobble back and forth. That's what we see on Mars: a very large wobble. [Earth] has only very subtle changes. That keeps the planet pretty stable relative to Mars." Venus, of course, doesn't fit this model, because it has no moons and is at least as stable as Earth. But that's probably because its much slower spin—once every 243 Earth days—may be slow enough to preclude wobbles, even without help. The price, however, is very long days, which on a world with a thinner atmosphere would produce enormously hot days and super-cold nights.

So, what of the Earth's future?

In the long run, the Sun is continuing to heat up, and eventually Earth's climate will go the way of Venus's. But there's no need to hurry building a spacefaring ark: the Sun probably won't get hot enough to boil off the oceans for a billion or more years. As for our own effects on climate, the good news is that in the million-year long run we're more likely to kill ourselves than ruin the climate as a whole. But that doesn't mean it's impossible to do both. "Both Venus and Mars seem to have gone from habitable to uninhabitable," Grinspoon says. "That should give one pause about messing with one's planetary climate without a complete understanding of the limits of that stability."

Author's Note

My favorite science meetings are often the ones no other science writers attend. That frees me from worrying about what BBC, CNN, or AP might be reporting on and allows me to focus on looking for interesting topics, knowing that whatever I find I'll be first.

Olympic athletes compete for gold medals. Reporters compete for scoops. This article is composed of a string of scoops, initially funded by Nature News and Cosmos, which trusted my belief that the Society of Environmental Toxicology and Chemistry (SETAC) held a meeting worth attending.

POISONS, TEMPERATURE, AND CLIMATE CHANGE

Will Global Warming Make Everything Else Worse?

Discussions of climate change generally focus on disasters. Warming glaciers that threaten to inundate coastal cities. Heat waves and droughts becoming ever longer, stronger. Hurricanes and tornadoes. Polar bears without icebergs. What they have in common is that they're all *big* events. After all, we're talking about *global* warming, *global* climate change. But we don't live on that scale. We live in our own homes, cities, counties, states. To the extent we feel the impacts of climate change, it's in our own backyards.

Consider a 2006 study in the *Proceedings of the National Academy of Sciences*. A team of climate scientists including Noah Diffenbaugh, then of Purdue University, broke the U.S. into a 4-

kilometer grid (much smaller than that used in global climate models) and examined the future of climate change on a single industry: winemaking.[76] Premium wine grapes like the weather warm, but not too warm. They are particularly sensitive to days when the temperature rises above 95°F, although extended stretches above 85°F aren't good for them either. What the scientists found is that while global warming wasn't projected to produce a massive change in average temperature in wine-growing regions (probably only about 5°F by the end of the century), that was enough to tip a lot of summer days into the heat-wave zone. For example, Diffenbaugh said, California's Napa and Sonoma wine regions currently see fewer than 14 of those 95°-plus days each year, but by the end of the century that could climb to 55 or 60—enough to pretty much put those regions out of business. All told, his team calculated, U.S. production of premium wines could drop 80 percent by 2100—even after accounting for the fact that some regions, including parts of Maine, Oregon, and Washington, could become California's heirs.

End of the world? Not really. Life without premium wines would be less interesting, but not unimaginable. Besides, America's loss might be Canada's gain. Or perhaps genetic engineering will come to California's rescue with heat-resistant strains. But it's the type of effect that hits us where we live—especially if that happens to be a place noted for finicky grapes.

Agriculture, obviously, isn't the only thing that might suffer. As alluded to earlier, there has been a lot of press about polar bears running out of the ice floes they use in their hunts for seals

---

[76] Michael A. White, et al, *Proceedings of the National Academy of Sciences*, 103(30), 11217-11222 (2006).

and other prey. Less known is a landmark study of rodents in California's Sierra Nevada. The study, which began producing results in 2006,[77] was a follow-up to a prior survey that ran an east/west transect across the Sierra Nevada in the 1910s, providing a rare snapshot of ecological conditions 90 years before. Amazingly, the follow-up found that many small mammals such as ground squirrels, voles, and mice have already shifted their ranges substantially upward, often by 1,300 feet or more. Presumably these animals are trying to avoid heat, but if warming continues there is a very real risk that some will run out of "up" in which to retreat. Other studies have found that changing climate is causing spring to come earlier and earlier in some areas, possibly shifting local ecologies out of synch with the migration patterns of birds, which are set by factors like the length of springtime daylight that won't shift with the climate.[78]

Synergistic Effects

As these studies show, climate change is likely to have a strong effect on ecosystems. But it won't occur in isolation. Sewage, industrial pollutants, agricultural runoff, automobile exhaust—these and more will continue to the affect the environment. "We need to look at climate change and other stressors and how they interact," says Nicholas Gard, an environmental toxicologist at the Bellevue, Washington office of Exponent, a large environmental and engineering consulting firm. "We really don't know what the effects of climate change

---

[77] Described in detail in Anthony D. Barnosky, *Heatstroke: Nature in an Age of Global Warming* (Island Press, 2009), pp. 69-82.
[78] See, *e.g.*, William Bradshaw & Christina Holzapfel, "Evolutionary Response to Rapid Climate Change," *Science*, 9 June 2006, pp. 1477-1478.

are going to be."

One factor that might cause old-enemy toxics to produce new problems is that changing climate might alter wind patterns, blowing smog into previously pristine areas. Increased floods, droughts, hurricanes, etc., might also stir such chemicals more thoroughly into the environment, and climate change could alter ecosystems, shifting the patterns of who eats whom. "As food webs change," Gard said at the 2010 meeting of the Society of Environmental Toxicology and Chemistry in Portland, Oregon, "animals and plants will become exposed to contaminants they haven't been exposed to before."

Consider, for example, the effect of climate on Arctic sea ice. The Arctic Ocean is one of the world's most remote bodies of water but it isn't immune to pollution. Plenty of long-lived pollutants, such as pesticide residues, eventually make their way into it from sources far to the south. Erin Mann, then a graduate student in environmental chemistry at the University of Toronto, Scarborough, Canada, has studied the interaction between several such chemicals, one of which was alpha-hexachlorocyclohexane, a degradation product of the banned pesticide lindane.[79]

Normally, sea ice keeps such pollutants in the water, sealing them off from the air. But as ice melts, more water is exposed to the atmosphere, allowing them to escape. "There's the possibility that global warming could produce more air pollution in the arctic," she says. Once in the air, these chemicals won't disperse back to the south. Rather, a meteorological phenomenon called

---

[79] Erin Mann, *Developing Methods for Studying the Fate and Transport of Contaminants in Snow and Ice,* masters thesis (2011).

the arctic "cold trap" will keep them in the north until they fall back to the surface in snow and rain. Some will return to the ocean where they came from, but some will wind up on land, contaminating previously pristine ecosystems and finding their way into human food. "[These chemicals] are bioaccumulative," Mann says, "so they'll increase as you go up the food chain. They are already at high enough levels that Inuit populations accumulate significant levels."

Sunburned Streambeds

Mann's findings involve a complex line of causation, but that's typical of the way many of the new findings operate. And not all are from remote areas. Will Clements is an ecotoxicologist at Colorado State University who is interested in the impact of global warming on summer runoff in mountain streams in the Colorado Rockies.

Water levels in these rivers (and many others in the American West) are heavily influenced by snowmelt, which typically peaks in the late spring. Warming, however, is expected to reduce winter snowpack—replacing much of it, in many regions, with rain. This will shift the patterns of runoff, increasing stream flows in the winter (or early spring) and reducing them in late spring and summer. Decreased summer stream flows are a concern throughout the West because they mean less water for both fish and agriculture (and possibly a need for new dams). But Clements wondered what the effect might be on damaging ultraviolet light reaching streambed organisms.[80] Partly, that's because water itself blocks ultraviolet light. But bank-full

---

[80] Clements, et al, *Global Change Biol.* 14: 2201-2214 (2008), and R.E. Zuellig, et al, *J. North Amer. Benthol. Soc.* 27:120-134 (2008).

streams, even the most pristine, are also siltier (or more turbid, in the argot of stream ecologists). Too much turbidity is a bad thing, but a little helps serve as an additional UV shield: sunscreen for riverbed organisms, if you will. And in Colorado, peak stream flows historically come in late May and early June, months when the sun is high in the sky, bathing the water in the greatest quantities of ultraviolet light.

To test the effect of potential changes in this, Clements' team placed UV screens over small segments of a dozen mountain creeks and left them for 60 days. At the end, they found dramatically higher populations of stoneflies, mayflies, and caddis flies in the screened segments than in adjacent unscreened ones. *I.e.,* reduced ultraviolet light was good for these flies, which form an important part of the streams' ecologies. Conversely, increased ultraviolet light (due to low stream flows) might be harmful.

But changes to stream-flow patterns are just the beginning. There is also increasing evidence that global warming can make toxic chemicals even more toxic. Biochemically, that's because a lot of processes occur more quickly—and therefore more dangerously—at higher temperatures. One of the simplest confirmations of this comes from Jennifer Goss, then a toxicology graduate student with the U.S. Army Corps of Engineers in Vicksburg, Mississippi, who studied the effect of climate change on lead toxicity (of interest to the military because of leaching from spent bullets on shooting ranges).[81] Her tests involved exposing *Daphnia magna*, a tiny freshwater crustacean commonly used in aquatic toxicity studies, to

[81] Goss presented her findings at the 2010 meeting of the Society of Environmental Toxicology and Chemistry.

dissolved lead at temperatures ranging from 68°F to 79°F. "We saw greater sensitivity at higher temperatures," she says.

Her team also found that at any given level of lead, higher temperatures caused the water fleas to reproduce sooner but die younger. What this means is that global warming might allow these short-lived crustaceans to evolve more rapidly, possibly helping them develop tolerances to environmental toxics—a potential mitigating factor.

That's an intriguing finding, because other ecologists have suggested that lifecycle length may strongly affect the ability of animals to adapt to rapidly changing climate. "Large animals like polar bears will probably do very poorly," says Christina Holzapfel, an evolutionary geneticist at the University of Oregon. "They have a long life cycle, so it takes them relatively long to adapt, genetically."

In other words, the future may belong to the fleas. Or at least to the ones who adapt most quickly at higher temperatures.

Big Effects

Because many effects occur in tandem, it can be difficult to pinpoint culprits for any given species' demise. Marjorie Brooks, an aquatic toxicologist from Southern Illinois University in Carbondale, Illinois, notes that in the past decade there have been tremendous declines in several types of fish in California's Sacramento/San Joaquin Delta (east of San Francisco Bay). "We were really befuddled by this," she says, "because [some] conditions are actually improving."

Gradually, her team realized that no single environmental factor was at fault. "We were seeing the compounding effect of multiple stressors," she says. "These animals are dying a death by

205

10,000 cuts, and one is water temperature." Increasing temperatures can affect aquatic organisms in three ways, she adds: (1) Warmer water carries less dissolved oxygen. (2) Most aquatic animals are "ectotherms," meaning their body temperatures are set by the environment. Warmer water increases their metabolic rates, producing higher need for oxygen, even as dissolved oxygen is dropping. What matters, Brooks says, isn't so much daytime temperatures as nighttime ones. "Animals can take a lot of daytime heat so long as water temperatures cool off at night," she says. (3) Thanks to the combination of sped-up metabolism and reduced oxygen, these animals must "breathe" more water for oxygen and consume more food, exposing them to ever-greater levels of pollutants.

It doesn't take a huge temperature increase to have a major effect, Brooks adds. In studies of fathead minnows (a commonly used laboratory species) the lethal level for dissolved copper changed so strongly with temperature that with the level of warming projected by 2060, it will only take half as much to kill them as it does today. "This has some pretty severe implications for water quality criteria," she says.

Not that the news is entirely bleak. For pyrethroid insecticides (commonly used in pet sprays, shampoos, and to kill mosquitoes that spread West Nile virus), Brooks says, toxicity actually declines with temperature. The same applies to neurotoxins such as DDT and mercury. For pyrethroids, it's because warmer temperatures speed up the biochemical processes that convert these compounds into nontoxic ones. For neurotoxins, it's because nerves are less sensitive at warm temperatures. "Try hitting your thumb with a hammer when your hands are cold," Brooks notes. "[It's] much more painful."

Of greatest concern, she says, are the potential effects on the smallest creeks and tributaries, which not only comprise 80 percent of the length of [all] rivers in the continental U.S., but whose shallow depth makes them most likely to vary with air temperature. They're also the places where many species breed, she notes, "and larvae are more susceptible to heat and contaminants."

It's a worry that tracks strongly with other scientists' concerns about the impact of climate change on other toxics, particularly pharmaceuticals, which reach wastewater by a number of routes, mostly when people flush unwanted drugs down the toilet or excrete them in their urine, hours after popping a pill.

Since we use these chemicals to cure disease we don't think of them as toxins. But if you're an otherwise-healthy fish, they're not a good thing. The same goes for trace chemicals from antibacterial soaps, sunscreens, or any of a host of similar products. Melissa Schultz, a chemist at The College of Wooster in Wooster, Ohio, has exposed fathead minnows to parts-per-trillion levels of triclosan and triclocarban: antibacterial agents commonly added to soaps, disinfectants, and (increasingly) such products as toys, bedding, socks, and trash bags. "These are levels we have found in the environment," she says, adding that while they are too little to produce visibly deformed fish, she has found that they affect fish behavior, particularly regarding mating.[82] In the wild, male fathead minnows aggressively defend the females' egg nests from rivals. But when exposed to triclosan and triclocarban, they are comparatively apathetic. "We are

---

[82] Schultz et al, *Environmental Contamination and Toxicology*, (1), 114-124 (2012).

finding subtle effects at environmentally relevant concentrations," Schultz says.

Dalma Martinovic, an environmental toxicologist from the University of St. Thomas in St. Paul, Minnesota, found something similar in zebrafish exposed to ibuprofen at 50 part-per-billion levels (only 2½ times higher than those sometimes measured in polluted rivers).

Ibuprofen and related anti-inflammatory drugs work by inhibiting the COX-1 and COX-2 enzymes. In arthritis patients, these enzymes are part of the pathway that produces prostaglandins, which play a role in inflammation. But in fish, prostaglandins help initiate sexual behavior, serving, among other things, as pheromones. Martinovic found significant reduction in courtship behaviors ("things like chasing and touching noses") in ibuprofen-exposed males. Furthermore, unexposed males showed reduced pheromone reaction to water collected from the tanks of ibuprofen-exposed females than from unexposed females.

Shultz's and Martinovic's work is pure toxicology. Ted Valenti, an aquatic toxicologist with the U.S. National Research Council (then on assignment to the federal Environmental Protection Agency in Duluth, Minnesota) has discovered that the effects of pharmaceuticals vary dramatically with the pH of the water in which they are dissolved—a factor that could change dramatically with climate change.[83] "That's the way these drugs are designed," he says, noting it helps the body route them to tissues where they are needed. But when the drugs find their

---

[83] T. W. Valenti, et al. *Integrated Environmental Assessment and Management*, Volume 7, Issue 4, pp. 636–647, October 2011.

way into wastewater, the same effect applies. "I've seen upward of ten to 20-fold differences in toxicity at pH 9 versus pH 6," Valenti says.

One of the factors affecting a river's pH is the photosynthetic activity of algae and streambed plants. During the day, these plants use dissolved carbon dioxide (a weak acid) for photosynthesis. That makes the water less acidic, raising pH. At night photosynthesis ceases, but metabolic processes continue, releasing carbon dioxide back to the water. The result: stream pH rises by day and falls at night. Not surprisingly, the degree of diurnal change is affected by stream flow, itself affected by climate. In a study of 22 tributaries of the Brazos River near Waco, Texas, Valenti found that the day/night fluctuation in pH was much greater in 2006 (an extremely dry year) than in 2007 (one of the wettest on record).

What this means in terms of ecological risk will depend on the chemical, but it's very likely that the most important times are those when pH climbs and the chemicals are at their most toxic. After all, if you're an organism exposed for 12 hours to a dose that can kill you, what difference is there if there's another 12 when you're exposed to a low dose?

Obviously, the import of Valenti's findings depends on the anticipated effects of global warming. In the parts of Texas he was studying, climate is expected to become drier. Also, millions of people in Dallas, Houston, Austin, and San Antonio will need more reservoirs to store water, magnifying the effect. And what all of this means is that on the local, ecological scale, the effects of climate change could be a lot more complex than most of us generally realize. That's not what we want, but "these are things we need to think about," says Exponent toxicologist Nicholas

Gard.

Author's Note

Science writers generally spend their time pitching ideas to editors, hoping to get assignments. But sometimes, the process goes the other way around. The next story wasn't even on my agenda until an editor at Cosmos asked me if I'd like to give it a whirl. The title comes from the fact that the initial story printed in 2013, when 2063 was fifty years off in the future. Since then, I've incorporated material from an Analog article that published three years later, but retained the original title.

ENERGY 2063

The Future of Global Energy Production

George Parrott is living in the energy future. In 2007 he built a state-of-the-art energy-efficient home in suburban Sacramento, California. A year later, he equipped it with a 3.7-kilowatt solar-cell array—easily enough to supply America's average residential electrical use of about 1,000 kilowatts per month. The year after that, he and his partner Chris Iwahashi added another 1.4 kilowatts and bought a pair of electric cars. They also installed another set of solar panels to provide hot water for showers, laundry, and dishes. On top of that, Parrott—not an engineer but a retired psychology professor from California State University in Sacramento—opted for an innovative solar-array design of his own imagining, using the panels as an awning to help shade his house from mid-summer heat.

The first array wiped out the couple's electrical bill. The second freed them from the gas pump. Both, he estimates, will

> "Sacramento has the fifth-worst air quality in the U.S. We believe that people who can afford it have an obligation to do as little damage…or as much repairing…to the environment as we can." – George Parrott.

pay for themselves in about 6.5 years—partly thanks to a California law that lets him sell home-generated power to the grid at market rate. That means that on summer days when people with less-efficient homes are dialing up the air conditioning, people like him can sell excess power at peak rates…then buy back in the evenings what they need for such off-peak uses as laundry or charging the cars, at as little as one-fifth the daytime price.

Parrott's and Iwahashi's lifestyle is a model for a decentralized energy future in which people generate their own energy, much as our ancestors once raised their own vegetables. But is it realistic?

Seattle, Washington futurist and science-fiction writer Brenda Cooper thinks it is and that such energy self-sufficiency is indeed the wave of the future. In part, she says, it will be driven by the desire to insulate ourselves from the worst effects of climate change. "Bigger and broader storm systems are knocking out our old [power] grid pretty easily," she says.

> "The world electricity market started with small decentralized power stations, near where the power was needed. I think we're going to go full circle. Everyone's rooftop could become a generator—not just of electricity but of hot water—things we use in the home." – Wes Stein

"More self-sufficiency [will] allow households and businesses to be free of the grid. In a climate-change challenged world, this will be seen as a smart move." But if we're striving for a carbon-neutral future to put the brakes on climate change, rooftop collectors may not be enough. Even people like Parrot and Iwahashi aren't completely off the grid. According to the U.S. Energy Information Administration, residential needs account for only 22% of American energy use. It takes additional energy to build their cars, run the offices where they work, pave the streets they drive on, and build energy-efficient homes.

By 2063, projections are that the world will have 9 billion people, all demanding developed-world standards of living, says Thomas Mason, Director of Oak Ridge National Laboratory, which carries out much of America's energy research. And while there is much to be gained from conservation, history says we find ever-new ways of using energy. "Our standard of living is defined by energy consumption," Mason says. "Standard of living equals energy use, and vice versa." From one point of view, this is a bleak thought: a prescription for war, poverty, and—if we continue relying on fossil fuels—global environmental collapse. But it's also a challenge. "That's why we need better alternatives," Mason says.

Happily, there are a number of alternatives, of which solar is one of the leading contenders. In a large, lightly populated country like Australia, for example,

> "If you imagine a 9 billion world population at the standard of living we enjoy in the developed world, you can't make the numbers add up in terms of [conventional] energy. Demand will exceed what we can support and the environmental consequences would be pretty infeasible." — Thomas Mason

the entire nation's electrical needs could be met with a perfectly efficient 50 x 50 kilometer solar array in the sunniest portions of the desert. "Land is not an issue," says Wes Stein, Solar Power Manager at CSIRO's National Solar Energy Centre in Newcastle, New South Wales. "There's certainly enough."

Solar-electric technology is a bigger concern. There are currently two basic technologies in use: photovoltaics—which generate electricity directly from light—and "concentrating solar power" (CSP), which uses mirrors to produce heat by focusing it. At the moment, photovoltaic cells are the primary form of solar power. And while they represent less than 0.1% of total global energy production, they're a proven technology that has been steadily dropping in price ever since NASA and other space agencies realized they were a good way to produce electricity for satellites and spacecraft.

For years, though, photovoltaics were viewed as too expensive for anything but specialized uses such as powering satellites or instruments at remote weather stations. But just as the computer industry boasts "Moore's Law"—which says that computing power doubles every 18 months—photovoltaic technology has "Swanson's Law." Named for Richard Swanson, founder of a California solar-panel manufacturer, this rule-of-thumb holds that solar-panel prices drop 20% with each doubling of the industry.

That's not yet enough to let everyone live like Parrott and Iwahashi—their home-energy system is affordable only because of the fortuitous mix of California sun, tax breaks, and the state's buy/sell law that allows them to take advantage of the difference between peak and off/peak electrical rates. "They're only able to do that because of subsidies of one form or another," says Alan Krupnick, an environmental and resource economist with Resources for the Future, a Washington, D.C. think tank. But the day is drawing nearer when solar might be able to stand on its own. According to tracking data collected by Bloomberg New Energy Finance,[84] solar-cell prices have fallen by a factor of more than 100 since 1997.

The other primary solar energy technology is CSP. The concept is simple...and ancient. As far back as 213 B.C., Archimedes reputedly used the same principle to set fire to invading warships by having the defenders of his city focus sunlight on them with hand-held mirrors. Modern attempts to replicate this have cast doubt on whether it was truly possible with hand-held mirrors, but the principle is well established. It's the same one used by scouts the world over to start a fire with a magnifying glass or construct a solar oven for baking bread. With a large-enough array of mirrors focused by computers, CSP technology is definitely capable of being ramped up to a large-enough scale to create giant powerplants capable of replacing their fossil-fuel counterparts. The only thing different is the source of heat.

The holdup comes from the combination of construction costs and thermodynamics. CSP "fuel"—sunlight—is free, but the high-performance mirrors used to concentrate it are expensive. Thus, to compete with fossil fuels, CSP plants need to be more efficient. That, in turn, requires temperatures at the focus to be quite hot. "Thermodynamic theory says that if you

---

[84] *As cited in:* "Pricing Sunshine," *The Economist*, 28 December 2012 (no byline), http://www.economist.com/blogs/graphicdetail/2012/12/daily-chart-19.

increase operating temperature, you increase efficiency," says Stein. In theory, this means that the answer is simple: just focus the heat more intensely. But it turns out that if you do this, things start to burn up like Archimedes' invading warships. Not literally, but chemically, as the heat takes a toll on turbine blades and pipes.

One solution is to wait for metallurgists to come up with alloys that can better withstand extreme conditions. Another is to make the turbine more efficient without changing the operating temperature—a boon not only to CSP but to existing fossil-fuel plants.

Conventional turbines use two "cycles" for tapping heat for electricity. In one cycle, steam turbines boil water, then use the expanding steam to spin the turbine blades. Unfortunately they lose large amounts of energy in the second "cycle," which converts the steam back to water. Another type of turbine, called a gas turbine, bypasses this by using hot gases that never have to be re-liquefied…but these lose comparable amounts of energy by having to re-compress the gas at the start of each cycle.

A newer alternative now starting to draw attention uses supercritical carbon dioxide instead of water or gas.[85] Supercritical fluids are substances normally thought of as gases, held at high temperatures and pressures that give them the characteristics of both a gas and a liquid. Since they don't have to be re-liquefied after each cycle there's no energy loss from that. Nor is there any need to waste energy by re-compressing them. The result might be a 25 to 35 percent increase in efficiency. "The coal, nuclear, and CSP industries are all starting

---

[85] Don't worry; the carbon dioxide is continuously reused, rather than being released into the atmosphere to contribute to global warming.

216

to look at that," Stein says—though the greatest benefit might be to CSP, where increased efficiency could do much to offset the cost of the mirrors. Shifting to supercritical carbon dioxide can also be done at temperatures already common for conventional steam or gas turbines. "You need only 700°C to 800°C and that's pretty doable with the metals available today," Stein says.

What will the energy world of 2063 look like? "Who knows?" says economist Krupnick. "There may be things we can't even envision." That said, he lists hydrogen fuel cells and improved batteries as important possibilities, the latter because it's "a possible game-changer." But it's just as likely the mix will be much like today's: including wind, nuclear, biomass and old-fashioned fossil fuel.

Wind is the current poster-child for alternative-energy success. Every summer day in America's Pacific Northwest, a parade of giant trucks heads out the freeway east of my home, carrying giant pylons and turbine blades into the sagebrush steppes of the eastern Columbia River Gorge, where half-mile-high bluffs funnel gale-force winds from the Pacific Ocean toward the inland desert. In Oregon, wind power already accounts for nearly 10% of electrical needs. Other places have done even better. In Spain, wind energy now supplies 15% of national electricity demand. In Denmark, it's 26%.

Another country with wind potential is Australia, though it currently lags at only 2%, according to Jonathan Whale, a wind-energy researcher at Murdoch University in Murdoch, Western Australia. But wind farms, he says, are going to be his nation's "go to" form of renewable energy if it plans on meeting its stated renewable energy target of 20% by 2020. Even greater contributions might come in the future, he says, if wind farms move offshore, where ever-larger windmills can be built with components brought in by barge. But wind has its disadvantages because, like solar, it's fickle.

There are two ways to deal with that. One is to interconnect numerous wind and solar plants onto a large grid, averaging out the fluctuations. But that can make one city's power dependent on another region's weather. For example, notes Mason, "If the wind changes dramatically in the Baltic [where much of Europe's wind power is generated] the Czech Republic sees their electric supply [become] unstable.."

The other solution is to find ways to store the energy for later use.

When most people think of storing electrical power, they think of batteries, but at gigawatt levels these would be enormous and absurdly expensive. A better way to do it, says Stein, is through "thermal storage"—a method that's particularly appealing for heat-based generators such as CSP facilities. This works by storing heat in well-insulated vaults of…well, anything hot. Ideally the material would also be dense, so it wouldn't require vats the size of supertankers to hold it. "Molten salt is an example," says Stein. "Nearly all of the CSP projects in Spain and the U.S. are integrating molten salt storage. When clouds come over, or it's nighttime, you use the heat to generate steam."

"When I first started researching wind energy, a 600-kilowatt turbine was considered large. Now the standard model is a couple of megawatts. In 2009, in Spain, I got stuck behind a semi-trailer carrying one wind-turbine blade. A single blade [can be] 82 meters long, just two meters shy of the length of the largest aircraft in the world." — Jonathon Whale

218

Photo-what? Novel Alternatives to Solar Panels

• *Thermoelectrics.* Nearly 200 years ago German physicist Thomas Johann Seebeck discovered that when different metals are joined, with one hot and the other cold, an electric current will flow between them. The process, long used to make temperature sensors called thermocouples, might be used to power cell phones, watches, medical sensors (possibly using the owner's body heat), or increase automobile efficiency by converting waste heat to electricity. "It's a well-known effect but hasn't really been used in the solar world," says Stein.

• *Thermionics.* Discovered by Thomas Edison in 1883, the thermionic effect is the "boiling away" of electrons from a hot cathode. Traditionally, it's been used to make current flow through vacuum tubes, making possible everything from pre-transistor radios to old-style television tubes. For these, electricity was used to heat the cathode, consuming it rather than generating it. But there's no reason solar power couldn't be used to heat the cathode, creating a thermionic generator.

• *Artificial photosynthesis.* Sometimes referred to as the "artificial leaf," this uses sunlight to break water into hydrogen and oxygen. The oxygen is released to the air, while the hydrogen is stored as fuel or used to power a fuel cell. The chemistry is complex, but more or less mimics that of natural photosynthesis, which also uses sunlight to split water into hydrogen and oxygen. "Whether you realize it or not, leaves are buzzing with electricity," says Daniel Nocera, a chemist at Massachusetts Institute of Technology. The process works well in tests. "We can put [an artificial leaf] in a bottle of water, hold it up to the sun, and see hydrogen and oxygen bubbles coming off," Nocera said at a meeting of the American Chemical Society in 2011. It also might be a way of making cheaper solar panels. "The price of the silicon in a solar panel isn't much," Nocera says. "A lot of the cost is the wiring. What this does is get rid of all that."

Similar chemical processes can be used to produce "solar fuels." There's a lot of talk in the environmental community about a "hydrogen economy" in which "clean, pure, hydrogen" will eventually replace dirty old gasoline. Hydrogen does have a major advantage in that it's pollution-free. When you burn it, the only byproduct is water. There's just one problem: hydrogen is a beast to handle. It's explosive. Its tiny molecules easily permeate out of storage containers. It can turn pipeline metals brittle. And to use it in a car, you need either to liquefy it or compress it, both of which take a lot of energy. Better, argues Ellen Stechel, deputy directory of Lightworks, a solar-power project at Arizona State University, is to take that hydrogen and produce something easier to use…such as solar-derived gasoline.

Stechel's work has grown out of an interest in finding ways to make synthetic aviation fuels, other than via biofuels (fuels derived from biomass), which cannot be generated in sufficient quantities if we still want to grow crops to eat.

Nobody, she argues, is going to make a commercially feasible electric jetliner in the near future. Even electric cars have their disadvantages. No matter how cheap batteries become, the ones we have don't recharge quickly and their weight reduces energy efficiency except in cars used solely for short-range commuting. Like it or not, conventional fuels have no such disadvantages. "The fuels we are using today have really nice properties," she says: "high energy density both by mass and volume." They're also extremely convenient. "We can fuel very fast, whether it's a big truck, an airplane, or a car," she says. "You can't do that with hydrogen and you can't to that with electricity." And they're fuels we're already well equipped to handle. "We have an enormous amount of infrastructure," she says.

There are a number of ways to make such fuels from air, water, and sunlight, but Stechel's main focus is via CSP. If the sunlight is focused on a metal oxide, such as zinc oxide or tin oxide, that oxide will give up its oxygen. When the reaction chamber is cooled back down and exposed to a mix of steam and carbon dioxide, the metal atoms will "steal back" the oxygen, producing not only the original metal oxide (ready for reuse), but carbon monoxide and hydrogen. "Those are the building blocks to make any hydrocarbon you like," Stechel says.

It's not a far-out technology. "We believe it could be pushed to market in about a decade," she says.

In the interim it might be possible to make solar-hybrid fuels. These start with a fossil-fuel source, then add solar energy to make that fuel even better. An example, says Stein, is using solar energy to split natural gas into syngas—the same mix of hydrogen and carbon monoxide Stechel's group is looking at making into pure solar fuels. The solar-energy proportion is only 20 or 30 percent, Stein says, but it's a starting point. "Without knowing it, [consumers] will be using solar energy."

More problematic is the role of nuclear power. "My personal feeling is that we are suffering from a massive case of 'Fukushima shock,'" says Krupnick, referencing the Japanese reactor that in the wake of the 2011 earthquake and tsunami produced the worst nuclear disaster since Chernobyl. It hasn't helped, he adds, that "fracking" technology, in which high-pressure steam is used to fracture rock strata and release oil and gas reserves previously believed to be unobtainable, has produced a new boom in fossil-fuel production. "With low natural gas prices, there's just no interest in nuclear," he says.

Mason believes the nuclear future (if we let it happen) belongs to "generation III" nuclear plants and small "modular" reactors. Small reactors are just that. They're placed underground and produce 50 to 300 megawatts of power, far less than the thousands of megawatts from current reactors—more like the reactors used on submarines and aircraft carriers. Because they're smaller and underground, they're safer, especially because they don't need to keep large supplies of fuel on hand.

Generation III reactors are designed to be more failsafe than their predecessors. For example, the cooling system might be in a gravity-fed reservoir on the roof of the containment building. Cooling water would flow into the reactor via gravity, then heat to steam that would rise back to the roof, where it would cool and condense back into water for the next cycle. The entire system is a continuous, passive loop relying only on the fact that water flows downhill and steam rises. "That's a passive safety feature," Mason says. "There is no need for operators or pumps."

Another improvement, Mason says, is the use of "accident-tolerant fuels." Old-style fuels, he says, were encapsulated in a material known as zircaloy, which had the advantage of not absorbing neutrons (an impediment to reactor performance) but which, when overheated, caused water to disassociate into hydrogen and oxygen—the source of the explosion that produced the Fukushima accident. New-generation fuels are encapsulated in materials that don't have this property. "If something happens and the system heats up, you don't compound the problem by having this release of hydrogen," Mason says. "It gives you a much greater safety margin."

Other approaches can massively reduce the quantities of waste while increasing fuel efficiency. One is fuel recycling. "In the current fuel cycle, it's only a couple percent of the potential energy that you've really used," Mason says. "That means the waste volume per unit energy is large. You can do a whole lot better." One option is the breeder reactor (known since the 1940s) in which uranium-238 (which composes 99% of natural uranium but isn't usable as reactor fuel) is converted to plutonium-239, a usable fuel. Another is the thorium fuel cycle, which converts thorium-232 into uranium-233, which doesn't occur in nature but is a perfectly usable fuel source. "Because thorium is more abundant than uranium, there's been renewed interest in this," Mason says.

With these methods, he says, it's possible to generate all of the world's electricity from nuclear sources. But he adds, "I'm not sure that would be valuable or desirable. There's a value in diversity of energy supply. I'm not sure you would want to put all your eggs in one basket." Nor does he expect fusion power to suddenly step up and fill the clean-energy gap, even in the next 50 years. "You might hope that by the end of that time frame you've answered the scientific questions and [are] beginning to answer the engineering problems," he says.

What this means is that the future will need a mix of energy facilities: centralized and decentralized, residential, commercial, and industrial. Few of the likely changes, however, will come quickly. Your cell phone might be obsolete in two years, your laptop in five, but power plants are vastly more expensive and designed to operate for decades. "The energy industry is very conservative and generally working with technologies that have

20-, 30-, or 40-plus-year asset lives," says Stein.

Mason agrees. "Fifty years sounds like a long time, when you think about healthcare and information technology. But in 'energy space' 50 years is not very long. If you look at the time it's taken for pervious transitions—nuclear power, petroleum, before that, coal—each took half a century, and in some cases longer. You're talking about capital investments that persist for 50-year time frames. It can't happen fast because the system has so much inertia in it."

If more rapid change does occur, Stein says, it will probably be driven by climate change. "We just had a fairly severe heat wave on the east coast of Australia," he says, "and we're getting significant, non-normal climate events around the world. If something did happen, where significant heating led to significant drought that lead to significant socionomic change, we could see a drastic acceleration in the move to these new technologies. But if it's business as usual, you'll probably wind up with a slow shift."

If changes do happen slowly our lives may not be all that radically affected. "Ultimately, I don't see that lifestyle should be affected," Stein says. "If we do these technologies correctly, the consumer should not see any difference when they turn the switch or start their car. They should still get their electricity. They should still get from A to B in their vehicle. There should not be any noticeable effect at the end of the day except that, perhaps, the climate is cleaner."

Author's Note

The next story came to me from a geology meeting but it's really about history. Or archaeology, with a touch of forensics. And horses.

Horses are cool in and of themselves, but if you're looking for a segue, they were also one of the ancient world's first power sources. Even today, we rate automobile engines in terms of horsepower. But really, this one is for the equestrians.

THE SEARCH FOR THE WORLD'S FIRST EQUESTRIANS

Once, there was a boy. Or perhaps a girl, though in the cultures of the time, a boy was more likely. We'll call him Khan, although the language groups from which that name would arise wouldn't evolve for thousands of years.

Khan was a hunter—perhaps a young one, still fresh with innovation. One day, as he, his father, uncles, and brothers stalked the thundering beasts of the plains, they cornered a mare in a box canyon, killing her with spears, arrows, and stones. But the mare had a foal that stood trembling nearby, torn between panic and the instinct to stay by its mother.

There were plenty of men to butcher the mare and carry the meat home, and as they set to work, Khan watched the foal. There, he realized, was more meat: not that anyone needed it at the moment. But if he could get it home alive, and if it was old enough to survive, he could keep it until meat was again needed.

Carefully, he fashioned strips of rawhide into a longer line. More carefully still, he approached the foal and wrapped the line around its neck, like he sometimes did with the dogs that lived in his village, sharing meat and fire and hunting with his clan. And slowly, cautiously, he led the foal home.

The foal lived and grew. Hunting was good, and so it was allowed to live, through winter and into spring and on to the next year. By this time, it was a pet—a source of meat only in the direst of emergencies. Then, Khan had another idea. The foal had become large and strong: far larger than the dogs which, as a child, he'd tried to sit on so they could carry him, laughing, around the village until the elders said dogs weren't made to be sat on and besides, they would bite if you played too rough.

One day, Khan led the young horse to a large rock. He climbed onto the rock, and from there, leapt to the horse's back. The animal flinched, but didn't buck or try to run. And the course of humanity shifted forever.

Okay, it probably didn't happen exactly like that. More likely, there were many Khans who independently made the same discovery. But someone had to be first, and the smart money says it probably happened on the Asian steppes, perhaps 6,000 years ago. Trying to prove that, though, has kept archaeologists busy for years.

Archaeology is one of the favorite fields of science fiction.

It's perfect for adventure stories: rife with mystery and shadowy stories of origins—a realm of vanished civilizations and priceless treasures. But real archaeology isn't *Indiana Jones*. Yes, the occasional Incan or Egyptian treasure surfaces. In 2005, for example, a team of American and Italian Egyptologists found pieces of the world's oldest seafaring ship, along with a 4,000-year old cave on the Red Sea, where Egyptian sailors had created a supply base, then mothballed it.[86] Inside, they found coils of rope and other supplies, neatly stacked for a return that never came. But such finds are headline-grabbing rarities. Rather than clambering around in teetering ruins, the real work of archaeology increasingly involves chemistry, remote sensing, and even laboratory experiments. Such, in fact, is the case with the search for the first signs of horse domestication. This article focuses on horses, but if we ever find traces of vanished civilizations on other planets, the odds are that the work of piecing them together will have a lot more in common with this than with *Tomb Raider*.

First of all, let's define what the horse researchers are looking for. Our friend Khan didn't domesticate anything. He caught a wild horse and tamed it. True domestication involves captive breeding. This turns out not to be easy. Horses are big, unruly animals, and stallions can be mean. Compared to raising a wild foal as a pet, breeding horses in captivity is a much more difficult task, says British researcher Marsha Levine of the McDonald Institute for Archaeology at the University of Cambridge. In the early days, she argues, it would have been simpler just to capture foals and tame them on an as-needed basis. In fact, she notes,

---

[86] See, among others, Andrew Curry, "Egypt's Ancient Fleet: Lost for Thousands of Years, Discovered in a Desolate Cave," *Discover*, June 2011.

only a few hundred years ago, America's Plains Indians still preferred stealing horses or capturing wild ones to breeding their own.

Genetic Clues

Obviously, there did come a time when horses were domesticated. One interesting line of evidence about the manner in which this came about can be found in the genes of modern horses (including today's "wild" ones, which are simply domestic escapees). A 2001 genetic study by biologists at UCLA and three Swedish universities, for example, examined their mitochondrial DNA, which is inherited from their mothers via the protoplasm of egg cells.[87] It found that modern horses are descended from a large number of wild mares, probably from diverse areas.

Initially, that was taken to mean that horse domestication arose independently in a number of cultures. Then a 2004 study in *Nature Genetics,* revealed that the male "Y" chromosomes came from a far more limited number of stallions.[88] This may mean that horses were domesticated by one single, clever tribe and that as the idea spread, others captured additional wild mares but mated them with domestic stallions obtained from their

---

[87] Mitochondria are cellular powerhouses that help "burn" food to provide energy. They are fairly self-contained subcellular features that have their own DNA. Mitochondrial DNA studies have also led to a somewhat controversial hypothesis that modern humans appear to be descended from one great-to-the-nth grandmother, sometimes called the "mitochondrial Eve." The work with horses is similar, but reaches the opposite conclusion.

[88] G. Lindgren, et al, *Nature Genetics* 36: 335–336 (2004). A subsequent study in *PLoS-ONE* paints a slightly more complex view, but reaches the same basic conclusion: genetic variations in modern horses' Y chromosomes is "extremely low." See Barbara Wallner, et al, *PLoS-ONE*, April 3, 1013.

neighbors. That certainly makes sense. Mares are easier to capture and tame, so if there are wild horses around, you might as well get your own, for free. But stallions are a different matter, and if domestic ones are already available, most people might rather trade for one than risk their lives trying to capture a wild one.

Sandra Olsen, an archaeologist at the Carnegie Museum of Natural History in Pittsburg, Pennsylvania, believes horses were first domesticated on the Eurasian steppes: that broad swath of grasslands that sweeps from Mongolia to Russia. And for more than a decade, she helped excavate villages from the Botai culture of what is now northern Kazakhstan, looking for evidence that 5,600 years ago, they were already relying heavily on horses for meat, tools, and transportation.

These villages are one of two sites vying for the honor of having the earliest signs of horse domestication (the other is a similar-aged village called Dereivka, in the Ukraine). And there is no doubt that horses played a major role in Botai culture. In one village, Olsen's team recovered more than 300,000 animal-bone fragments. "Of those, over 99 percent are from horses," she says. Nor has her team found any indication the Botai engaged in agriculture. "It was a horse-centric society," she says.

Much of the debate over horse domestication is a polite point/counterpoint between Olsen and Levine. Regarding the Botai diet, for example, Levine counters that the villagers may also have eaten fish. "We don't know that they were subsisting entirely on horses," she says. "Fish bones don't preserve well, archaeologically."

Still, horses were obviously important to the villagers. (The same is true for Dereivka, which also contains a startling number of horse bones.) Cut marks on the bones indicate that the horses were butchered for meat, and Olsen thinks there is no way the villagers could have killed that many horses simply by hunting. "You have to imagine what conditions were like if they were hunting wild animals on foot," she says. "The villages had 160 or more houses, so how would they sustain themselves and not deplete the herds immediately?"

Levine, on the other hand, makes exactly the same argument *against* domestication. How, she asks, could the villagers graze that many horses nearby without depleting the grasslands? "There is a tendency among people who study horse bones to say there are lots of horses in this site, [so] that means they're domesticated," she says. "But all it means is that lots of horses died and the bones were brought there." In fact, she says, if you go to more recent sites where horses were known to be domesticated, you don't tend to find such large numbers of bones. "When you find large numbers [of horse bones], that tends to be an indirect indication that they're probably wild," she argues.

Indirect Evidence

One might think it would be easy to figure out if the villagers were raising horses for slaughter rather than hunting them. After all, domestic horses need equipment such as bridles, lead ropes, and hobbles. But unfortunately these would have been made from leather thongs, which would long ago have rotted away. It's also perfectly possible to ride a horse without the use of either a saddle or bridle. "During the early stages of horse domestication," Levine says, "it is likely that they were usually ridden that way." Olsen agrees. If you look at classical Greek

232

statues of people on horseback, she says, "they don't have saddles, horseshoes or stirrups. And that's in 400 B.C.E. [only 2,400 years ago]."

The earliest surviving pieces of horse tack were probably "cheekpieces" from bridles: fasteners used to help tie together the various thongs of the bridle, near the horse's cheek. Some archaeologists think that odd pieces of worked antler found at Dereivka are cheekpieces, but the artifacts are sufficiently nondescript that Levine dismisses this as wishful thinking. They could be virtually anything, she says. Instead, the earliest unambiguous evidence of horse domestication doesn't come from much before 2000 B.C.E., an era from which archaeologists have found not only many clearly recognizable cheekpieces, but two-wheeled carts that look like chariots, buried not only with their wealthy human owners but with the horses that presumably pulled them.

Chariots are an advanced technology. By the time they appeared, Olsen says, horse domestication was "a done deal; you know you're missing the boat." Levine concurs: horses didn't instantly go from being wild animals to being buried in graves with chariots. "They have to have been domesticated at that stage for some period of time," she says.

With other domestic animals, like dogs and cats, one way to look for traces of domestication is in bone-structure changes resulting from domestic breeding. After all, a modern Chihuahua doesn't look much like the wolf-like dogs that were its distant ancestors. But horse skeletons don't start to show much sign of this prior to 1200 B.C.E.—although a 2012 study of the genes responsible for coat color may have pushed the date for human breeding back as far as 3000 B.C.E.[89]

Another approach is by looking at ancient horse teeth for signs of bit wear, or at the skeletons of aging horses for abnormalities associated with being ridden or worked in other ways. Unfortunately, such evidence is hard to find. Worse, says Levine, the studies are conducted by archaeologists, not veterinarians, and tend not to be rigorous enough to rule out alternative explanations. "They're interesting," she says, "but I don't think they're very conclusive."

"It's frustrating," Olsen admits. "At the same time it's really fun: a great detective story."

Part of her own work has been at the Botai village of Krasnyi Yar, where her team discovered a circular array of ancient postholes that looks suspiciously like a corral. That discovery is itself an example of modern archaeology in action. Rather than digging up wide swaths of land, archaeologists can now conduct surveys with ground-penetrating radar and other instruments designed to find subtle variations in the soil's electrical resistivity and magnetic properties. The techniques, collectively referred to as geophysical archaeology, have been used to survey everything from Roman bathhouses, Viking longhouses, and Native American burial mounds to German bunkers destroyed in the D-Day invasion of Normandy.

At Krasnyi Yar, these techniques revealed 54 pit houses and dozens of postholes. Although the postholes had long ago refilled with dirt, they were visible to the imaging techniques because the fill had a different geophysical signature than the surrounding soil...and they formed suspicious corral-like

---

[89] Robin Bendrey, *World Archaeology*, 44(1), 135-157, February 22, 2012 (online).

arrangements.

Olsen was thrilled. Her work on the steppes has led her to believe that the best way to figure out how the ancients raised their horses is by watching what modern cultures do in similar circumstances—and in 1944 the Soviets produced a golden opportunity for this by forcing horse-herding Kazakhs onto communes: year-round villages not all that different from those inhabited by the Botai. "What [the Kazakhs] do is probably analogous to what they did in ancient Kazakhstan," Olsen says. And what they do is to corral their horses near their villages at night, then take them out to pasture during the day—just as the archaeology suggests for the ancient villagers of Krasnyi Yar.

Dung Heap

Finding the circular structures was exciting enough. But then the scientists had the brainstorm of looking at the paddock soil for traces of ancient horse manure. What they found were phosphates, which are one of the primary nutrients contained in animal manure. At a 2006 meeting of the Geological Society of America, Andrew Stiff, a graduate student with the project, reported that phosphate levels in soil samples taken from within one of the paddocks were ten times higher than those in the adjacent soil.

High phosphate levels, however, aren't proof of manure; they can also be created by cooking fires. To rule that out, Stiff's team examined another mineral, potassium, which should also have been elevated if hearth fires had been the cause. It wasn't, indicating that the find indeed represents manure.

Phosphate, though, isn't the only nutrient contained in

manure. Manure is also high in nitrates…and these were not elevated within the paddock. But that's actually good news, says Rosemary Capo, a professor at the University of Pittsburgh who was also a member of the research team. That's because nitrates easily leach out of soil in rainstorms or are decomposed by bacteria. Phosphates can remain for millennia. Thus, the lack of nitrates indicates that the phosphates aren't recent contaminants from a later corral built on the same site. "It suggests we've got old stuff," Capo says.

Other archaeological evidence indicates that the horses were raised in the village rather than being hunted in the surrounding wildlands. One such line of evidence comes from the fact that complete horse skeletons, including skulls and vertebrae, have been found in the villages. Hunters, Olsen argues, wouldn't have bothered to bring back these heavy, useless bones. "We call it the 'schlep effect,'" she says. Instead, they would have butchered their kills in the field and carried back only the parts they needed. Furthermore, archaeologists haven't found many arrowheads at the site, something that should have been plentiful in a hunting culture. And when the scientists used the latest analytical tests to check for ancient blood residues on the arrowheads they did find, Olsen says, what blood they found was human.

Also revealing are the ages of the animals that were slaughtered. "Between 30 and 50 percent were killed young, which indicates culling," Olsen says. "That's standard in horse domestication."

Geological Society of America

Levine, not surprisingly, has a different interpretation. The horses, she says, appear to have died at ages roughly proportional to what would be found in a living herd, "as if a catastrophe took down the whole herd at once. This is typical of random hunting or herd driving [in the hunt... If you're raising them for meat, you kill them around age three or four]." Perhaps, she says, the circular structure was a holding pen for wild animals corralled for subsequent slaughter. "American Indians used to chase [wild] horses into a corral," she says, although she concedes that doing so in the middle of a village "might have been tricky." Or perhaps the corral was used for a few animals that, like our friend Khan's foal, were captured wild, then tamed. Among the vast numbers of horse bones found in the villages, it would be difficult to distinguish a few tame horses

237

from thousands of wild ones.

Similar concerns have been raised about the horse bones at Dereivka. There, Levine says, the most common age was seven to eight years. "That's not a time when you would butcher animals for meat," she says. But if Dereivka's hunters were selectively stalking wild animals in their prime, that's exactly the age distribution you might expect.

Why would hunters focus on the animals most able to put up a good fight, rather than the young, old, or lame? One clue, Levine reported in a 1998 study in the journal *Antiquity*,[90] comes from the fact that horsemeat is much higher in healthy, polyunsaturated fats than is meat from cows and sheep. The ancients wouldn't have known about polyunsaturated fats, but they might have known that horsemeat is healthier to eat—and in particular that its fat was a more easily digestible baby food than other kinds of fat. "When I interviewed [modern] Kazakhs, I found that horse fat and camel hump were the most desirable for weaning food," she says, adding that they are often given to infants mixed with pasta or simply cut into small lumps for them to suck on. Thus, if you're hunting horses, she says, "what you want is a big, fat male."

Not to mention that if you're stalking a herd of horses, the stallion will turn to attack, protecting its mares. "The result is that the humans will kill the stallion first," she says. "And if you're hunting for meat, that's all you need."

Her conclusion: the Botai used hunting tactics that killed off entire herds, while the people of Dereivka stalked individual animals, favoring stallions in the prime of life. There is no need,

---

[90] Marsha A. Levine, *Antiquity*, 72, number 275, pages 90–100 (1998).

she says, to presume that either group raised domestic horses for slaughter.

Olsen, however, has other evidence to back up her belief that the Botai had mastered horsemanship. One clue comes from the manner in which they made their stone tools. Without beasts of burden, it's a lot more convenient to do so at the flint quarry, rather than lugging large hunks of rock back home. "But they brought in big chunks and made tools [at the village]" she says. "That indicates that they had packhorses." Also, one of the most common tools found in the villages are thong-smoothers, used to make rawhide thongs, such as would be used in bridles and hobbles. Even more common are hide-scrapers—also needed to make leather thongs.

Still, none of this is proof. "We're talking about lots of different types of indirect evidence," Levine says. "It doesn't prove domestication; it's just interesting and possibly points in that direction." Olsen, though, thinks it's significant that all of the evidence points in the same direction. When you put it all together, she says, the simplest explanation is that the Botai's horses were domesticated.

Bye-bye, Indiana Jones—Hello, "CSI"

Additional secrets may also be hiding in the soil of the ancient animal pen. Only a few days before the geology meeting at which Andrew Stiff announced the phosphate finds, for example, Olsen learned that the soil was ten times saltier than that outside the pen. The probable source: urine, presumably from horses. In addition, she told me that her team was looking for additional evidence that the manure was truly from horses by hunting for traces of fatty chemicals unique to horse manure. "If we find those, that really nails it," she said.

Not that this would answer the question of whether the corralled horses were indeed domesticated. But chemists are also seeking an answer for that by analyzing Botai pottery for traces of fats from mare's milk—still a staple on the Eurasian steppes. Finding such fatty traces, Olsen says, might be as close to a "smoking gun" for horse domestication as it's possible to find. "You don't want to milk a wild mare," she says. (And on this point, Levine agrees. "I don't think you can milk a wild horse," she admits.)

And by 2009 it looked as though Olsen's team had found their smoking gun, publishing their results in *Science*.[91] But the evidence may not be as indisputable as hoped for. In a 2012 book chapter, Levine challenged the chemistry, arguing that while the fats found in the pottery fragments do appear to be equestrian, it's not clear that they are from *milk*. The finding is interesting, she wrote, but requires additional study to confirm it.[92]

Another step would be to conduct isotope analyses of trace elements such as strontium in the horse's teeth to determine where the animals were born and how far they ranged. This work is still in its early stages, Olsen says, but "We are finding indications that one horse seems to have been born in the south and maybe traded into the village . . . We're reconstructing their lifestyle and finding out a lot about these people."

But even if someday it is conclusively proven that the Botai

---

[91] Alan K. Outram et al, *Science*, 323:1332-1335, March 6, 2009.

[92] Marsha A. Levine, "Domestication of the Horse," in Neil Asher Silberman (ed.), *The Oxford Companion to Archaeology*, 2nd ed., Oxford Univ. Press (2012), 15-19.

villagers of Krasnyi Yar indeed raised horses, this doesn't necessarily mean that they were the first. "What is the chance that you find the first of anything?" Olsen asks.

Meanwhile, archaeological methods continue to evolve. Classic tales of crawling through underground tunnels have their roots in older techniques, epitomized by century-old finds in Egypt, such as King Tutankhamen's tomb. Today's methods may sometimes invoke such drama, as with the Egyptian caves, but much of the newer work is becoming more and more sophisticated, involving delicate tests for teasing indirect evidence out of the most mundane sources. "Archaeology is heading [toward] a lot more physics and chemistry," Olsen says. The trend is so strong, in fact, that future archaeologists may lament how their predecessors cavalierly threw away the best materials from their digs.

And what might these be? "The soil," Olsen says.

If she's right, astro-archaeologists of the future won't be tromping willy-nilly over any ancient sites they might find in their explorations. More likely, they'll be scared to even walk on them, fearful of destroying trace evidence. Sadly for those who like the drama of crumbling edifices, trapdoors, and secret passageways, archaeology of the future may be more like crime-scene investigation than rambunctious old Indiana Jones.

Author's Note

If you're into geography, Missouri has some odd city names. There's Butts, and Knob Lick, Strain, Tea, Tightwad, and Peculiar. But it's also dotted with locations named for other parts of the world, with the map revealing towns like Japan, Cuba, New London, Venice, and Berlin.

I'm not sure when I first heard of the New Madrid Seismic Zone, but one of the first things I did learn about it was that it's pronounced MAD-rid, not Ma-DRID because…well, why is Cairo, Illinois, pronounced KAY-row? The people who settled those areas were an independent-minded sort.

But a couple of hundred years ago, they also had some really big earthquakes.

SHAKE, RATTLE, AND ROLL

Is Missouri Really America's Most Dangerous Earthquake Zone?

At 2:15 a.m. on December 16, 1811, residents of what is now the northeast corner of Arkansas were shaken from their beds by a large earthquake, powerful enough to knock even today's houses out of plumb and shift them off their foundations. In the next two months there were three other big jolts and countless smaller ones, spreading damage across pieces of what are now five states (Arkansas, Missouri, Tennessee, Kentucky, and Illinois).

In 1811-12, however, the region was so sparsely populated that damage was relatively light, and it's likely the death toll was zero. There were no seismometers to measure the exact size of the earthquakes. Time passed, and memories faded. Cities were built with no thought that they might be sitting on a seismic time bomb. Earthquakes, everyone knew, happened in California, not

Missouri. Then, in the latter half of the Twentieth Century, scientists and historians rediscovered the disaster and realized that these quakes—collectively known as the New Madrid quakes for the town of New Madrid, Missouri (pronounced MAD-rid) —were among the strongest ever recorded in the Lower 48 states—at least as strong as the 1989 Loma Prieta earthquake, which knocked down freeway spans in the San Francisco Bay Area, killed 63 people and injured 3,700 more.

Meanwhile, geologists were finding evidence that the Nineteenth Century earthquakes hadn't been unique. For at least 2,000 years, the New Madrid Fault Zone had popped loose big ones about every half-millennium. New Madrid is only 200 kilometers south of St. Louis. Suddenly, the Gateway City wasn't looking like the safest place to have built a tall, freestanding arch.

The first formal estimates of the earthquakes' magnitudes came in 1973, from the late Otto Nuttli of St. Louis University, who looked at historical descriptions and concluded that all four quakes must have had magnitudes between about 7.0 and 7.5.[93] By way of comparison, the official U.S. Geological Survey estimate of the magnitude of San Francisco's devastating 1906 Earthquake is 7.9. In 1985, Arch Johnston of the University of Memphis upped the magnitude estimate to 8.0 or larger, based in part on reports that the quake reversed the flow of the Mississippi River and rang church bells as far away as Boston, Massachusetts.[94] Suddenly, the New Madrid quakes were the

---

[93] O. W. Nuttli, *Bulletin of the Seismological Society of America*, Vol. 63, No. 1, pp. 227-248, February 1973.

[94] A. C. Johnston & S. J. Nava, *J. Geophysical Research*, Vol. 90, No. 88, pp. 6737-6753 (1985).

largest ever recorded in the contiguous U.S. states.

Johnston's wasn't the biggest estimate. A video called "Hidden Fury," produced for use in geology classes, describes the New Madrid earthquakes as the "most powerful series of earthquakes ever known on earth."[95] "[The earthquakes] grew in the telling," says Seth Stein, a geophysicist at Northwestern University whose book about New Madrid, *Disaster Deferred*, was published in September 2010. "This was a good-sized earthquake, [the Earth] has had lots of bigger ones."

In 2000, Susan Hough, a seismologist at the U.S. Geological Survey's Pasadena, California office, took a new look at the historical records. "You have an account that says people were frightened and ran outside, and chimneys came down," she says. "It's all breathless, but the bottom line may be that it was just a couple of chimneys. The older the accounts and the more fragmentary, the easier it is to exaggerate."

She and a collaborator also discovered that there were no contemporaneous newspaper references to Boston church bells ringing in response to the earthquake. Perhaps the newspapers didn't bother to mention it, she says, "but that makes no sense. The newspapers did mention that the quake was felt in other places." She thinks the church-bell story may have arisen from confusion between Charleston, South Carolina—300 miles closer to the epicenter—and the Charlestown section of Boston. Once established, however, the story became a major part of New Madrid lore: in Hough's words, "a myth that refused to die."

Similarly, reports of the Mississippi flowing backward may

---

[95] www.bullfrogfilms.com/catalog/hid.html.

simply have involved a seiche (sloshing) caused when banks caved in. But they could also be accurate, even if the quake wasn't anywhere close to magnitude 8. "I'm not so sure the river 'flowing backwards' should be discounted," says Stein. "The reporters were experienced boatsmen." One account, collected by the Historical Society of Missouri, reads:

> The current of the Mississippi was driven back upon its source with the greatest velocity for several hours in consequence of the elevation of its bed. But this noble river was not to be stayed in its course. Its accumulated waters came booming on, and over topping the barrier thus suddenly raised, carried everything before them with resistless power.

This doesn't sound like a seiche. But did the earthquake really elevate the entire riverbed, forcing the water to run upstream? Stein thinks it might simply have been a big eddy created by short-lived dams formed by vertical slippage of the Reelfoot Fault (one of those involved in the earthquakes), which crosses the Mississippi three times. When powerful currents spill over such dams, the water plunges deep, with the majority of the current hugging the river bottom. This, in turn, creates suction—well known to river rafters—at the base of the dam. The suction is strong enough to cause the surface water to flow backward, giving the appearance that the river as a whole is running upstream.

I myself grew up along a sizable river in Illinois where at high water the backward-flowing surface current created by our town's eight-foot dam could extend a quarter-mile downstream. On a river the size of the Mississippi, the effect could be much larger, lasting until the earthquake-produced dam eroded away. And it doesn't require a San-Francisco-level earthquake to produce it. In a 2000 paper in the *Journal of Geophysical Research*,[96] Hough concluded that the earthquakes hadn't been anywhere close to magnitude 8.0. Rather, she decided, Nuttli's original estimates had been right, with the strongest jolts weighing in somewhere around magnitude 7.5.

Art by user Soronk on de.wikipedia [Public domain], via Wikimedia Commons

A lot of other scientists agreed. "I've always felt the early estimates were dramatically over-estimated," says Andrew Newman, a geophysicist at Georgia Institute of Technology.

---

[96] Hough, et al, "*J. Geophysical Research*, Vol. 105, pp. 23,839-23,864 (2000).

One problem with the older estimates, he says, is that the New Madrid Fault Zone is relatively small, only about 150 miles long, and big earthquakes are produced only when large chunks of the Earth's crust slip. "There is no way you really can fit a magnitude 8 earthquake in such a small seismic zone," he says. "Even a 7.5 could have a bit of trouble."

Another problem for the older, giant-earthquake idea comes from GPS measurements of long-term deformation of the rocks on opposite sides of the fault.

In standard tectonic theory, the forces that act on the crust to produce earthquakes are continuous, but earthquakes themselves occur in sudden jolts, when faults stick, then slip, releasing accumulated strain. Between earthquakes, the crust bows, as pressure mounts. The rate at which this occurs is a measure of how rapidly the fault is picking up strain, and the best GPS instruments allow this to be measured with remarkable precision. Rocks on each side of California's San Andreas Fault, for example, are bowing north/south at the rate of about 35 millimeters per year. That's enough to produce a 4-meter slip— enough for a magnitude 8 earthquake, roughly every 100 years. Thus, if New Madrid is popping loose a magnitude 8 every 500 years or so, you'd expect it to be accumulating strain at about one-fifth the rate of the San Andreas. But that's not the case. Instead, says Eric Calais, a geophysicist at Purdue University, the rocks on each side of the fault are barely moving, instead building up strain at a rate of no more than 0.2 millimeters per year (less than an inch per century). And, he adds, that's an upper boundary. "The actual number could be zero."

Unless there's something odd about the New Madrid Fault Zone, this may mean the fault has mysteriously shut off. Stein thinks that's indeed what has happened. In a 2009 paper in *Nature*, he and Mian Liu, a geophysicist at the University of Missouri, Columbia, argued that even though the New Madrid seismic zone continues to generate small earthquakes, they are simply long-delayed aftershocks from the big ones in 1811-1812.[97] In the big, rapidly shifting fault zones of Chile, California, or Indonesia, Stein and Liu argue, aftershocks generally tail off within a matter of weeks or months. But in mid-continent, everything occurs much more slowly. "In the middle of a continent, aftershocks can go on for hundreds of years," Stein says.

Alternatively, the earthquakes were a lot smaller than magnitude 8—smaller, even, than 7.5. In 2010, Hough reduced her estimates yet again, presenting them at a meeting of the Seismological Society of America in Portland, Oregon. Rather than relying solely on her own expertise, this time she asked experts in four earthquake-prone countries—Canada, Italy, India, and the United States—to review the historical accounts. Helping with the interpretation were data from the U.S. Geological Survey's "Did You Feel It?" web page, which invites people to submit reports of their experiences of current earthquakes whose magnitudes can be measured by seismometers. Her new conclusion: the New Madrid earthquakes were "right around" magnitude 7. "Possibly a bit below, possibly a bit above. But not as big as 7.5."

Not that a magnitude 7 is something to be sneered at. "Haiti was a magnitude 7," Hough notes.

Her new estimates are within the realm of possibility for fault

---

[97] Seth Stein & Mian Liu, *Nature*, 462, 5 November 2009, pp. 87-89.

zones of New Madrid's size, but they still don't jive with the GPS data. If mid-continent faults behave like plate-boundary faults—a big "if," Calais is quick to point out—0.2 mm of strain accumulation per year isn't enough to produce even magnitude 7.0s at 500-year intervals. Rather, he says, the largest earthquakes that can be produced that frequently by this rate of strain accumulation are on the order of magnitude 6.[98] "If the type of earthquake that tends to repeat on the New Madrid Fault were a 7.0, then at 0.2 mm per year we would get one every 5,000 years," he says. One prospect, of course, is that Hough's estimates are still too high and that the earthquakes were magnitude 6, not magnitude 7. But before we arbitrarily downgrade them yet another notch, let's look at the underlying tectonics.

Cracking the Heartland

The big question is why there are any earthquakes in the New Madrid area at all. New Madrid is smack-dab in the middle of a continental plate, and traditional theory says earthquakes occur at continental margins, where plates are bashing together, pulling apart, or grinding sideways against each other. The heartland of a continent is supposed to be stable.

The short answer is that nobody knows for sure. But continents aren't monolithic blocks. They've been sutured together over the eons, as plates bash into each other, stick together, then break apart again. Liu adds that the New Madrid Fault Zone dates back about 700 million years, to the breakup of

---

[98] This calculation, which he did during a telephone conversation with me, is based on a chart in an article he and Stein published in *Science*, Vol. 323, No. 5920, p. 1442, March 13, 2009.

an ancient supercontinent called Rodinia (from the Russian word for "motherland"). He calls the area a scar in the Earth's crust. "The general notion is that these rift zones are weak places that tend to cause stress concentration," he says, though he adds, "We know that is not all of the story because we have many of these ancient rifts. Some are seismic, such as the one in New Madrid, but others have no earthquakes at all."

So, what makes New Madrid special? One prospect is that it's simply the weakest spot in the mid-continent—a place where stresses will naturally focus. But there's no geophysical evidence that this is the case, says Stein. If it were a weak spot, he says, the underlying rocks would be warmer than the rest of the continent, because hotter rocks are weaker. "But we can't find a thermal difference."

Furthermore, there is geological evidence that New Madrid isn't the only sometimes-active seismic zone in the Midwest. There appear to be several, with earthquake activity periodically shifting from one to another. In Southern Indiana, for example, the Wabash Valley Fault Zone appears to have produced a few big earthquakes 6,000 years ago. A thousand years ago, there was earthquake activity in Oklahoma. There's also an old fault zone in Mississippi. New Madrid itself only became active a few thousand years ago; before that, it was quiescent for millions of years. "These things switch on and off with time," Stein says.

In part, these shiftings of stresses may be facilitated by the simple fact that tectonic forces in the mid-continent are lower than at plate margins. That gives the crust more time to make subtle adjustments between earthquakes. But it also means that there's more opportunity for local factors to play a role. For example, Stein says, at New Madrid some of the triggers may have been changes in the Mississippi River that shifted sediments from one location to another. This would have altered the weight distribution of the sediments, possibly helping to unlock the fault zone. Similar changes elsewhere might have made other faults less likely to slip. "It's a complicated system of interacting things," Stein says. "It really is like the butterfly that flaps its wings in Brazil and causes a tornado in Texas."

Thousands of Feet of Ice

All of this, however, begs the question of where the stress originates. Hough thinks it's glacial rebound. The idea is simple: during the Ice Age, much of the northern part of the continent was covered by thousands of feet of ice. The weight depressed the crust, causing it to sag into the squishier rocks of the mantle. Then the ice melted and the crust started rising back to its pre-glacial elevation. In places, the rebound has been dramatic. Beach terraces in Manitoba show nearly 1,000 feet of rise in the past 8,000 years.[99] Even today, the Hudson Bay shore is rising at about half an inch per year. Farther south, the crust behaves like a teeter-totter pivoting on a "hinge line" at about the latitude of the Great Lakes: north of the hinge line it is rising, and south of it, it is sinking—though if you go far enough south, the effect vanishes.

This up-and-down pivot produces stresses. In the north, there's little doubt that they've triggered some big quakes. A map

---

[99] John P. Bluemle, "Glacial Rebound, Warped Beaches and the Thickness of the Glaciers in North Dakota," www.dmr.nd.gov/ndgs/ndnotes/Rebound/Glacial%20Rebound.htm.

assembled by Natural Resources Canada[100] shows many, in a line across southern Ontario and Quebec and on into the Maritime Provinces and Labrador. The biggest, magnitude 7.2, struck offshore from Newfoundland in 1929, producing a tsunami that killed 28 people.

In the New Madrid Fault Zone, the evidence for glacial-rebound triggering is less clear. On the pro side is the fact that something switched the fault on not long after the ice sheets melted. "The timing is good," says Georgia Tech's Newman, "but we don't really have a smoking gun." On the con side, New Madrid (and Canada) aren't the only places to have had geologically recent earthquakes. In 1886, Charleston, South Carolina saw a magnitude 7.3—roughly the size of Hough's re-estimated magnitudes for the New Madrid temblors. In 2006, the Gulf of Mexico was hit by a 6.0. Both are too far from the one-time glaciers for rebound to be a likely cause.

Not that all quakes have to result from the same forces. But Stein also doesn't believe glacial rebound is occurring at New Madrid. "The GPS data that show glacial rebound beautifully to the north don't show anything happening at New Madrid," he says.

Newman thinks that overall, the type of motion seen in Charleston, New Madrid, the Gulf, and the hundreds of smaller quakes that hit the East and Midwest each year show a different story, making it look as though the U.S. is being squeezed from the east and west. There's even an obvious source of pressure. On the east, it's the spreading of the ocean along the mid-Atlantic Ridge, a force that drives Europe eastward and North

---

[100] See: atlas.nrcan.gc.ca/site/english/maps/environment/naturalhazards/ earthquakes/majorearthquakes.

America to the west. On the other side of the continent, it's the crumpling that's created the Rocky Mountains. There's a tendency to think of continents as rigid sheets, but nothing that big is perfectly rigid. Squeeze it hard enough, and it will crack.

Under this explanation, it's possible that the reason there's no significant buildup of strain across the New Madrid Fault Zone is that Stein is right and the fault has shut off. But that would mean the strain must be accumulating somewhere else. Geophysicists are currently looking for where it's accumulating, via a couple of thousand GPS stations scattered across other Midwestern and Eastern fault zones. But so far they've seen nothing, says Calais, though that may simply indicate that even more monitoring stations are needed. At present, he estimates, only about half of the region is adequately monitored and many GPS stations are designed for use by surveyors, lacking the precision needed to spot slow-moving tectonic processes.

But, though the debate between glacial rebound and shifting continent-wide tectonic stresses may still be unresolved, at least we have two viable theories for the production of large, mid-continent earthquakes without the rapid buildup of GPS-measurable strain. In fact, says Calais, the strain might be so delocalized that it will be difficult to detect even with the best GPS instruments. Meanwhile, the folks in the Midwest have to live with the uncertainty.

More Dangerous Than California?

When I was growing up in Illinois, about a four-hour drive from St. Louis, I'd never heard of the New Madrid Fault Zone. Then, one summer we got a small earthquake—too small to do damage, but enough to rattle dishes and set the entire town

254

abuzz in a manner usually reserved for near misses by tornadoes. By the time I graduated from college, I'd heard of New Madrid, heard the story about Boston church bells, and "knew" I'd grown up near the "most dangerous earthquake zone" in the country.

And so, officially, it remains. "Right now, on the books, 'hazard' for New Madrid is higher than California," says Zhenming Wang, a seismologist and geotechnical engineer with the Kentucky Geological Survey. "Clearly, compared to the science, [this] doesn't make sense." There are a number of reasons why the hazard has been overrated, he says, ranging from technical to quasi-political. From the technical perspective, for example, Wang says that the current hazard maps use a very conservative parameter for the rate of attenuation of ground shaking with distance from the epicenter, substantially widening the distance at which severe damage might occur. In the quasi-political realm, hazard managers have chosen to base estimates on a 2,500-year time horizon, rather than the 500-year time horizon used in other areas. There's no right or wrong in this choice, but the longer the time horizon, the bigger the disaster you have to guard against.

Add to this the increasing evidence that the largest historic earthquakes were actually magnitude 7 rather than magnitude 8, and the whole thing snowballs into building codes that might be excessively restrictive. "That is very unfortunate," says Wang, who notes that people in Paducah, Kentucky can't build two-story houses. Not all of the New Madrid area is so strongly affected, he adds. For example, he says, Memphis, Tennessee has chosen not to base its building codes on the super-high-hazard maps. "But in Paducah, they are in a terrible situation."

At the same time, hazard elsewhere might be underestimated. Newman thinks the findings about shifting seismicity and continent-wide "squeezing" mean that faults hundreds of miles from New Madrid might someday cut loose with big temblors. "I thinks the hazard is much more distributed than what we've previously shown," he says.

Still, a large earthquake near St. Louis, Memphis, or anywhere in the vicinity would be a big deal. "A magnitude 7 is still pretty dangerous," Newman says. "It's just not the same level of danger as a magnitude 8 or a magnitude 7.5."

Author's Note

The first time I went to Iceland, I'd never heard the term "bucket list." I just realized it was a place I'd always wanted to go, and that I wasn't getting any younger. I took my bicycle, pedaled most of the circumference of the island, and fell in love with the place, even though the weather...leaves a lot to be desired.

The second time I went, a couple of years later, I realized I was "visiting" Mars.

PALEOLAKES, JÖKULHLAUPS, AND MOBERGS

What Iceland Reveals About "Wet Mars"

August in Reykjavik, Iceland is a mix of the familiar and exotic. On the one hand is the usual crush of vacation-minded folk preparing to hit the great outdoors before summer breathes its last. On the other is a climate in which "summer" resembles April in Seattle, Washington.

In Icelandic tourism, the name of the game is "rain shadow"—an issue of being on the sunny side of the island, away from the prevailing storm track. If you're Icelandic or rich enough to afford some of the word's priciest car rentals, you drive. Budget-minded visitors ride buses, which means you become very familiar with the nation's primary bus hub in downtown Reykjavik.

I was on a hiking vacation with a few days to kill between major treks when I found myself in that bus station, trying to decide if the weather was telling me to go south to visit Keiko the whale (from the 1993 movie *Free Willy*) or head the opposite direction to the Snæfellsness peninsula, the jumping-off point for Jules Verne's *Journey to the Center of the Earth*. Stalling, I was reading travel brochures when I stumbled across a poster listing upcoming cultural events in Reykjavik. To my amazement, one was an international conference on Mars polar science and exploration.

Perhaps because Jules Verne was already on my mind, I was instantly intrigued. Try as I might, I couldn't come up with a reason why the conference should be held in such a pricey destination. Eventually, I shrugged, visited Jules Verne country (in sunshine), and followed that with a weeklong trek through the starkly beautiful highlands of central Iceland. I didn't know it at the time, but by traveling, I'd found my answer. At the end of my Icelandic vacation, when I begged my way into the conference as a science writer, I learned that Iceland may be one of the closest earthly analogs to prehistoric Mars. The scientists, many of them Icelandophiles like myself, flock here to see in real life the landforms that tantalize them on satellite photos.

We've all seen the headlines. Between photos from orbiting cameras and finds by ever-more-sophisticated rovers, it appears that the Red Planet was once scoured by substantial quantities of water: flowing, pooling, and fanning out to sculpt canyons, lakeshores, and deltas. But how "wet" is "wet" remains an open question. In 2000, Michael Malin and Kenneth Edgett of San Diego - based Malin Space Science Systems, Inc. released now-famous photos showing scalloped cliffs that looked as though they had been undercut by seepage from a line of springs—a truly stunning finding.[101]

You don't need steamy temperatures, however, to produce such features—not, at least, if you have snow, ice, sun, and not absurdly low temperatures. Cold-climate dwellers on Earth know this intuitively: their rivers don't freeze to the bottom during even the most bitter cold snaps. But the best evidence comes from the Dry Valleys of Antarctica, where there is no chance that the liquid might simply be left over from warmer days.

The Dry Valleys are a freeze-dried desert that hasn't thawed for thousands or millions of years. They're "dry" because they're largely ice-free, but they do have water, melting out of mountain glaciers that extend partway down the peaks that hem them in. The heat source that produces this liquid water is the sun. On clear days, sunlight penetrates the top layer of ice, where it is absorbed by the relatively opaque boundaries between crystals. Because ice is a fairly good insulator, enough heat accumulates at these points to cause melting. Microscopic rivulets trickle downward until they emerge from beneath the glacier as a modest-sized river. The effect is magnified by reflecting-oven conditions on the sunward sides of ice blocks or other terrain irregularities.

Today, there's not a lot of meltwater, but researchers have found traces of vanished "paleolakes" from the last ice age. Although that was a time when the Dry Valleys were even colder than today, the rate of melting back then appears to have been much higher than at present: shorelines have been found from an ancient lake that was 1,700 feet deep and perhaps 80 square miles in area. Why would a colder climate produce a bigger lake?

---

[101] Malin & Edgett, *Science*, 30 June 2000 pages 2330-2335.

In one of the presentations I attended in that conference in Reykjavik, B. L. Hall of the University of Maine posited that the colder Paleolithic climate might have reduced the number of storms by creating enough extra pack ice to force the nearest open water several hundred miles farther offshore. Fewer storms would mean more sun and more melting, despite the lower temperature.

If Antarctica can generate big lakes during an ice age, then similar things might have happened on Mars. Streams might have gushed from beneath glaciers, cut canyons, and flowed into icy oceans, even if the air temperature never rose above freezing. This would require more water than we see today, but most of it could have cycled from glaciers to ice-covered oceans, with only a small amount flowing freely at any time.

Noah's Lake

So far, we've generated lakes and streams, but not floods. On Earth, these are generally associated with hurricanes, monsoons, and other events that spread a lot of moisture over a large area. It's highly unlikely that Mars was ever *that* wet. But we can produce a big flood with less water by finding a way to drain a lake in a hurry. One way is if the lake was created when a glacier surged across the path of a stream. Such ice dams are highly unstable (largely because if enough water builds up behind them, they start to *float*) and they can give way catastrophically, producing huge floods (see Chapter 5). And at least one satellite photo has given a hint that Mars may once have had surging glaciers.

What the photo shows is a kinked *esker*.[102] Eskers originate as a river channels beneath glaciers. With time, the channels fill with debris melted out of the glaciers, so that when the ice melts, the streambeds become snaky ridges. On Earth, the largest of these, in Mauritania, is 500 feet high, but smaller ones are common in parts of the American Midwest, where they trace the melting of the great continental ice sheets.[103]

Martian landforms that look like eskers are one of the better indicators that the planet once had glaciers extending beyond its present ice caps. But a kinked esker is even more intriguing, looking like a snake with a broken back. On first impression, that's baffling. But such eskers show up—where else?— in Iceland, associated with glaciers that have been known to surge and retreat, folding older eskers into accordion-like pleats called *concertina eskers*.

Fire + Ice = Jøkulhlaup

Surging glaciers damming modest-sized lakes are all we need for a flood big enough to catch the attention of a Martian Noah. But eskers, paleolakes, and glacial surges might actually be a red herring. We don't need them to produce Martian floods; we don't even need warm enough temperatures for sunlight to melt the ice. That's because Mars once had a much better source of heat. Mars had volcanoes. And that, more than anything else, is what brings Mars scientists to Iceland.

As its name indicates, about 12 percent of Iceland is covered

---

[102] J. Nussbaumer, et. al, *Second International Conference on Mars Polar Science and Exploration*, abstracts, p. 137 (2000).

[103] The one in Mauritania formed hundreds of millions of years ago, when Africa was near the South Pole.

by ice, mostly in large icecaps that rise against the horizon like surreal mounds of congealed fog. But the entire island is volcanic, much of it very actively so—and in places, the actively volcanic and the icy overlap.

Dramatic things happen when volcanoes meet glaciers. If hot ash falls onto one from above, the ice melts quickly and sluices downslope in a destructive mudslide called a lahar.[104] But the really spectacular events occur when an eruption occurs *beneath* the ice. In Iceland, this has happened often enough that there's even a word for it: *jökulhlaup*.

Subglacial eruptions begin gently. As the lava wells up, ice melts and turns to steam that melts more ice—but at first, the steam, lava, and water are all trapped beneath 1,000 or more feet of glacier. Seismometer needles may jiggle, but a person standing directly on top of a minor eruption might not even notice it. In a big eruption, the glacier might fracture into a vent that spits steam and ash onto the snow above, but often the only visible indicator is a gradually expanding depression in the ice cap, as layer after layer melts from below. On a steep slope, the water flows away beneath the glacier. But beneath a large, flatter ice cap, water is trapped in a subglacial lake. Eventually, the dammed up meltwater finds a way out. Typically, it does so in an enormous rush—the jökulhlaup.

Small jökulhlaups are local phenomena, like cloudburst-caused flash floods. Big ones are awesome spectacles, as glaciers gush forth torrents of water the color of chocolate milk with sufficient force to roll 15-foot boulders downstream. Iceland's most feared source of jökulhlaups is the volcano Katla, which

---

[104] Some Martian volcanoes show signs of mudflows that look like lahars.

lies beneath Myrdalsjökull, a 250-square-mile mantle of ice that reaches thicknesses of up to 3,000 feet. The largest recent outburst, however, came from beneath a bigger but less active ice cap called Vatnajökull. This flood, on November 5, 1996, peaked at a flow of 1.8 million cubic feet per second, giving Iceland the brief honor of having the world's second largest river and converting a span of the nation's longest bridge into a pile of twisted metal. The worst outflows from Myrdalsjökull, of which there have been several in recorded history, were probably six times larger.

In previous centuries, untold numbers of Viking and Icelandic farmers died in jökulhlaups, but modern seismology has given scientists the tools necessary to issue warnings to jökulhlaups-prone regions. These regions are easy to identify, and therein lies their relevance to Mars.

Characteristically, jökulhlaups rearrange landscapes in two major ways. On steep slopes, they carve canyons—flowing so fast and hard that they can even rip apart bedrock. Iceland's most dramatic canyon is Jökulsargljufur, a 20-mile long trench averaging 1,600 feet wide and 300 feet deep. Geologists believe it was carved in a mere one or two days by the largest known jökulhlaup, which blasted out of Vatnajökull about 2,500 years ago.

On flatter terrain, such floods slow down, spread out, and drop their loads of debris, building substantial deltas, Icelanders call sandurs—remarkably smooth plains in a land that is otherwise one of the most mountainous on Earth. Sandurs can be inland or coastal, occurring wherever a jökulhlaup loses momentum. They can also be quite extensive; Myrdalsandur (below Myrdalsjökull) covers more than 250 square miles and is by no means Iceland's largest. And just as a single flood can carve a canyon, a single outburst can build or substantially enlarge a sandur. For example, the last major outburst from Katla, in 1918, shoved the coastline more than half a mile farther out to sea.

Mars has canyons aplenty—and it has deltas that might be sandurs. Does this mean it has also had jökulhlaups? Nobody can say for sure, but it's certainly one of the more interesting wet-Mars hypotheses. One of its advantages, as Icelandic experience indicates, is that a lot of erosion can be done in a single event. A "fossil" glacier sitting atop a dormant volcano (something we'll get to in a moment) is all we need for substantial erosion, even in a gradually drying Mars. And given the vast erosive force of a jökulhlaup, the event need only happen once in any given location; we don't need a Martian climate that is wet (snowy) enough to support repeated cycles of glacier formation and flood.

What evidence is there that Martian floods were actually jökulhlaups? One indication was found near the 1997 Mars Pathfinder landing site, which shows all the signs of catastrophic flooding but has streambeds peppered with mysterious craters, up to about 300 feet in diameter. A group of researchers from the U.S., the U.K., and Iceland found a possible explanation for these craters by watching the outflow from a "small" outburst from the Myrdalsjökull two years after the Pathfinder landing.[105] The researchers observed large blocks of ice washing out from the glacier, later to be stranded as the flood receded. On Earth, when such ice blocks melt, they leave behind depressions called

---

[105] James W. Rice, Jr., et al, Mars polar science conference abstracts, p. 146.

kettle holes. Myrdalsandur is liberally strewn with kettle holes similar to the craters found at the Pathfinder site. If the Martian landforms are also kettle holes, that would imply that the Pathfinder site was once swept by floods bursting from beneath a vanished ice cap.

Questing for Maars

So far we've looked at the effect of subglacial eruption on the overlying ice, but not at the effect of ice (or water) on lava. But it can have several distinctive forms indicting the presence of long-gone water.

When magma erupts beneath relatively thin ice (or shallow water), for example, the water turns instantly to steam, producing a blast like a meteor impact and leaving a crater to match. Such craters are called maars, from the Latin word *mare*, for ocean.

Sometimes, such an eruption spends itself in the initial blast. A few miles east of the Cascade Mountains in Oregon, for example, a maar formed in this manner bears the highly descriptive name Hole in the Ground. A mile across, perfectly circular, and several hundred feet deep, it lies in nearly flat terrain, with such an inconspicuous rim that you barely notice the crater until you've practically stumbled into it. On other occasions, the volcano continues to erupt. Sometimes, that builds a conventional volcanic cone atop the maar, obscuring it. But often, the lava simply builds up a rim of spatter around the original pit. Hawaii's famous Diamond Head is a maar that appears to have been built up in this manner.

Iceland, not surprisingly, has many maars. But it also had a related type of landform called pseudocraters, most notably around its largest lake, Myvatn. Pseudocraters form when lava erupts on land, then flows across marshy terrain (or into a lake, across waterlogged sediments). The water trapped beneath the lava heats until it boils into superheated steam, then explodes, producing a flock of circular depressions (which at Myvatn range from 10 to 1,000 feet in diameter). Similar features could presumably form if lava flows across a glacier, or shallow permafrost.

A third possibility is that lava erupts beneath thick ice, where the weight of the overlying glacier prevents a steam explosion. Instead, the glacier traps the lava, cooling its outside and preventing it from flowing sideways. Depending on the shape of the volcanic vent, the lava then forms two characteristic features, *hyaloclastic ridges* and features Icelanders call *moberg* mountains (sometimes called *tuyas* by geologists from other lands). Hyaloclastic ridges are long, narrow, and sharply peaked. Mobergs are sheer-sided, looking like they've been formed by Jell-O molds. Iceland's most famous is Herthubreith, which rises above a sandur desert to an elevation of 5,500 feet. If a moberg pokes through the top of the glacier that crowned it, it often wears a more conventionally sloped volcanic crown, like a conical party-hat.

Possibility number four is that the lava erupts beneath deep water, rather than thick ice. In this case, the water again confines the lava's spread and forces it to cool rapidly. But instead of mobergs and hyaloclastic ridges, what you get are *pillow lavas*, characterized by piles of bulbous shapes like oversized pillows. Other rapid-cooling conditions (generally caused by water) can produce *columnar basalts*, in which the lava cracks into hexagonal columns such as those seen in Ireland's Giant's Causeway or

California's Devils' Postpile National Monument.

The final possibility is that a volcano erupts beneath "fossil" ice trapped in the soil either as permafrost or a buried glacier. The resulting landform isn't all that different from what you'd get with a traditional maar, but the idea of such buried ice deposits is particularly appealing on Mars, where the planet's one-time surface water seems to have vanished to somewhere (see Chapter 10), and underground is one likely explanation.

Earthly permafrost is generally shallow and not thick enough to supply enough moisture to make anything interesting happen. But again, Iceland is an exception. Its most recent buried ice deposit is in the vicinity of a volcanic crater called Askja. Now a major tourist attraction for those willing to brave a 12-hour, 4-wheel-drive bus ride, Askja blasted out Iceland's deepest lake (700 feet) in 1875. The midwinter eruption also blew 0.5 cubic miles of ash and cinders into the stratosphere and onto the surrounding landscape, burying (and insulating) enough snow that 7-foot layers could still be photographed in 1939, 64 years later. Slightly older buried ice was exposed by the 1996 Vatnajökull jökulhlaup, and ice estimated to have been buried more than 600 years ago has been found in other parts of Iceland.[106] The message is simple: even in a "warm" earthly climate, buried ice can persist for a long time. On Mars, "long" might be measured in millions of years, giving plenty of time for a subsequent eruption to find a buried glacier and release its waters in a canyon-and-delta-carving torrent.

In the years since my trip to Iceland, three rovers, Spirit, Opportunity, and Curiosity, have spent a combined time of

---

[106] J. Helgason, Mars polar science conference abstracts, p. 72.

nearly 20 years driving across Mars, sampling rocks, taking close-up views of geological formations, and producing overwhelming evidence that the Martian surface was once wetter than at present. But all that these rovers can produce are snapshots of the terrain they've visited. The big picture: how *much* water there was, and how widely distributed, is still, in large part, based on scanning the surface for maars, pseudocraters, mobergs, hyaloclastic ridges, pillow lavas, and columnar basalts. And some have indeed been found…probably. Moberg-like formations, for example, have been spotted in Valles Marineris, an enormous Martian trench that dwarfs any valley on Earth.[107] These mountains indicate that the giant valley may once have been at least partially ice-filled. Maars, on the other hand, are difficult to distinguish from ordinary volcanic craters, and a swarm of pseudocraters looks a lot like the bombardment from an asteroid that broke up shortly before impact. Still, recent studies based on meter-resolution orbital photos make it look ever more likely that Martian magma did indeed encounter water or ice, with Icelandic-style results:[108] enough so, in fact, that I myself would be very, very surprised if many of them didn't pan out as true lava/ice interactions.

Spiders, Fans, and CO_2 Geysers

Not all of Mars's Icelandic connections feature ice and lava. A quite different connection surfaced in 2000, when satellite photos revealed transient features near the south polar ice cap

---

[107] Mary Chapman, ed., *The Geology of Mars, Evidence from Earth-Based Analogs* (Cambridge University Press (2007).

[108] L.P. Keszthelyi, et al, *Icarus* 205, 2011-2029 (2010).

that looked like tangled knots of dark-hued rope. Closer examination revealed multi-limbed, vaguely radial patterns that earned the moniker "black spiders."

Surface features like that have never been seen on Earth, but Hugh Kieffer, a planetary scientist from the U.S. Geological Survey in Flagstaff, Arizona, suggests they might have something in common with another Icelandic feature: geysers. Or more specifically, geyser-like steam vents called fumaroles. Though instead of spouting volcanic gases like earthly fumaroles, the Martian ones may be conduits for hurricane-force jets of carbon-dioxide gas.

In the Martian winter, the poles are cold enough for carbon dioxide to freeze out of the air and deposit as frost, several feet thick. In fact, scientists estimate that each winter, fully one-third of the Martian atmosphere (mostly carbon dioxide) does this—only to evaporate the following spring. The mind boggles at trying to imagine what the Earth would be like if our own atmosphere did the same thing.

The Martian spring isn't warm by human terms, but it certainly is by the standards of frozen carbon dioxide: Kieffer's team calculated that there's enough heat to vaporize up to 20 pounds of carbon dioxide per square yard per day.[109] Furthermore, carbon-dioxide ice on Mars is transparent enough that the greatest heating comes at its base, where sunlight hits dark, underlying rock. And this bottom-up heating even occurs if the ice is contaminated by dust, because the dust motes will be the first things to be warmed. That will cause them to melt their

[109] Hugh H. Kieffer, et al, *Nature* 442, 793-796 (17 August 2006).

way to the bottom of the ice, leaving microscopic tunnels that quickly re-fill with dust-free carbon-dioxide ice.

Soon, carbon dioxide gas is building up beneath the ice…and that gas has to go somewhere. Kieffer predicts that small channels will feed into larger ones, like merging raindrops on a windowpane. Where the biggest channels break through to the surface, he calculates that the gas might roar out at speeds in excess of 120 miles per hour. Another prediction is that dust carried with the jets would deposit atop the ice in "dark fans" extending hundreds of yards downwind—exactly what has been seen in other high-resolution photos.[110]

The black spiders aren't true fumaroles and they're not true geysers, but they've got elements of each. And while the continuing operation of rovers will provide ever more information directly from our neighboring planet, it's unlikely they will do anything to diminish Mars scientists' fascination with places like Iceland.

Similar landforms speak of similar geological processes…and even if the parallels aren't always perfect, where else can you fly into a modern airport, check into a luxury hotel, rent a car, and drive to terrain that resembles the Martian poles?

Author's Note

At the same time I was visiting Iceland, I was also a regular contributor to Psychology Today. I didn't write the next story for them, but it definitely drew on that experience.

---

[110] See R. A. Lovett, "'Spiders' Channel Mars Polar Ice Cap," *Science*, Vol 289, 15 September 2000, pp. 1853-54.

THE TRANSIENCE OF MEMORY

We Really *Can* Remember it for You Wholesale

The 1990 movie *Total Recall*, based on the Philip K. Dick novelette "We Can Remember it for You Wholesale," envisions a future in which memories can be implanted well enough for you to "remember" a vacation (in the story, a trip to Mars) without actually taking it. As with other tales of hypnosis, brainwashing, and dream worlds, the story's power comes from the questions it raises about truth, fantasy, and the reliability of memory. Such stories all ask the same chilling question: What would it be like if we couldn't distinguish true memories from false ones?

The answer isn't as deeply in the realm of science fiction as we might like. Memory is notoriously slippery, and events we'd swear we witnessed might not have happened the way we remember. They might not have happened at all. This is a particular problem for childhood memories—even important ones which, when recovered in adulthood, may form the basis of long-after-the-fact accusations for crimes such as child abuse. How do we know whether these memories are accurate?

Consider the case of Donald Thompson, an Australian psychologist who, in the 1970s, was accused of rape. The evidence against him was very simple: the victim vividly remembered *him* as her attacker and could easily have picked him out of a lineup. But her memory *had* to be distorted, because Thompson had an ironclad alibi. At the time of the attack he was on live television—talking, ironically, about memory and memory distortion. The victim, it turned out, had been watching the program when she was attacked and had merged Thompson's face into her memory of the event.[111]

According to Daniel L. Schacter, former chair of the Department of Psychology at Harvard University, there are seven basic types of memory imperfections. In his book, *The Seven Sins of Memory: How the Mind Forgets and Remembers* (Houghton Mifflin, 2001) Schacter catalogs them as transience, persistence, absentmindedness, blocking, bias, misattribution, and suggestibility.

The first four are related to forgetfulness.

Of these, transience is the simplest. When I was in grade school I could recite the capitals of all 50 states and list most U.S. presidents, with dates. Today, I'd botch up quite a few. Where did that information go?

Even relatively recent memories suffer from transience. Try the following experiment: engage a friend in conversation, taping

---

[111] Jeremy Dean, *PsyBlog: Understanding Your Mind*, 5 February 2008.

it for later reference. Whenever your friend says something pithy, write it down, trying for a direct quote. If you're like most people, you'll be lucky to get beyond the 15th word before you start diverging from what's on the tape. Even trained reporters usually veer into paraphrase by word 25, and with each successive word the paraphrase gets worse. It's a humbling experience: you *know* what was said—you can almost hear it echoing around your brain—but it slips away faster than you can scribble it down. Worse, it slips away faster than you think it does, which is why it's easy for well-intentioned reporters to misquote people.

Absentmindedness and blocking are also common. In absentmindedness, you might tune me out for a paragraph and can't remember what I said. Or you promise your spouse you'll mow the lawn...and really *do* mean to do it. In blocking, you know something and *know* you know it—it's on the tip of your tongue—but you just can't quite find it. Blocking is disconcerting because it increases with age, making you fear you're getting Alzheimer's, but it's normal, even if not well understood.

But our main topic now is memory distortion, which Schacter puts in his last three categories: consistency bias, misattribution, and suggestibility.

Consistency bias is the interpretation of a memory in light of later knowledge. I coach distance runners, for example, who might find it hard to remember what it felt like to train for a race at which they eventually excelled. Were they confident or nervous? Tired or excited? Memories of confidence and relaxation could be colored by the knowledge that their training actually worked. There have been a number of good studies in this field, nicely summarized by British psychologist Jeremy Dean on his website PsyBlog. What they have in common is that whether it's remembering one-time political views, former romances, or pretty much anything else, we're likely to skew our view of how we felt about them *then* by how we feel about them *now*.[112]

Misattribution, on the other hand, is what nearly put Thompson in jail for a crime he didn't commit. Again, we all do this to some degree. A year from now, you may remember reading about this, but you may think someone else wrote it or forget whether it was in a book, *Scientific American*, or a friend's Facebook post. Misattribution is a big deal when it leads to erroneous criminal charges but merely a nuisance when it causes you to waste time looking in the wrong places for citations.

Suggestibility is by far the most interesting memory distortion. It's nothing less than the implanting of false memories—whether by accident or design: memories so vivid people would swear on a stack of Bibles they're accurate, sometimes even when they're frankly impossible. There are several ways to distort memory in this manner. In now-classic experiments, for example, Schacter used simple word tests in which people were asked to study a list of words, centering on a "theme" word not actually presented.[113] For example, the list might be_*candy, sour, sugar, bitter, good, taste, tooth, nice, honey, soda, chocolate, heart, cake, eat, pie*. Afterward, study participants were tested on their memory by being asked to identify words as

---

[112] *PsyBlog: Understanding Your Mind*, 20 February 2008.
[113] Susan A. Clancy, et al, *J. Abnormal Psychology*, Vol. 111, No. 3, 455–461 (2002).

"old" if they were on the original list, or "new" if they weren't. With the above list, most people will accurately identify something like *horse* as new. But many will be sure *sweet* was on it, when actually it wasn't. The reason: *sweet* is the theme word around which all of the others centered, and the list is designed to get you thinking about sweetness without actually using that word. Schacter views this as a type of misattribution: you yourself are thinking about sweetness and misattribute your own thoughts to the list. But it's also a form of implanted memory in which the experimenter carefully leads you down the garden path to a tempting but false conclusion.

This, of course, is a far cry from fabricating a trip to Mars. But other researchers have experimented with considerably more detailed memories. In 1995 Michelle Leichtman, a researcher in the Department of Psychology at the University of New Hampshire, conducted another now-classic experiment in which kindergarten and pre-school children received a visit from a man introduced simply as Sam Stone.[114] Sam's visit was very straightforward. He walked into the classrooms, looked around, and left, without saying or doing anything else. But beforehand, different classes had been told different stories about what to expect. Some had been told nothing other than the simple fact that Sam would visit. Others received detailed depictions of him as bumbling but kind, with stories of funny things Sam had done on other occasions. Five weeks later, the children were asked to recall Sam's visit. When asked nothing but open-ended questions, those who'd not been given the bumbling-but-kind stereotype responded with accurate memories. But some of

---

[114] Michelle Leichtman, *Developmental Psychology*, Vol. 31, No. 4, 568-78 (1995).

those who'd been prepped to expect clumsiness produced memories in line with that expectation.

Another group of children, including some who'd been prepped to expect clumsiness and some who weren't, were guided through their memories of Sam's visit with inaccurate, leading questions such as: "Do you remember when Sam accidentally tore a page out of the book? What was he wearing when he did that?" A few of the control-group children (those who'd not been told what to expect) fell for the trick. But among those who'd been told to expect clumsiness so many remembered such false incidents that Leichtman calls the implications "alarming" for the use of poorly recovered childhood memories as testimony in sex abuse cases. Overall, she says, leading questions, when combined with previously implanted suggestions, can powerfully distort children's memories. Worse, the children's reports of their now-false memories can be compelling, detailed, and remarkably durable.

Take Me to Your Leader

Children aren't the only ones who can suffer from this. Schacter's colleague Richard McNally, also professor of psychology at Harvard University, argues that many people claiming UFO abduction experiences may have experienced something very similar in adulthood.[115] The process is more complex, he says, but the end result is that memories so deeply engrained that these people may even suffer post-traumatic stress symptoms similar to those experienced by Vietnam veterans or sex-abuse victims.

---

[115] Richard J. McNally, *J. Experimental Psychopathology* 3(1): 2-16 (2012).

McNally began his study by running an ad in *The Boston Globe* asking, "Have you been abducted by aliens?" After screening out jokesters and those who thought he was talking about foreigners, he wound up with six women and four men, average age 48, who convinced him they truly believed they'd been abducted. On psychological profiles, these people looked perfectly ordinary except for a few signs of post-traumatic stress. They weren't depressed, anxious, or in any way psychologically abnormal. They did, however, test high in "absorption," a personality trait that allows you to become deeply engrossed in whatever has caught your attention, such as the imaginary world of a good novel. They also showed "a whiff" of *magical ideation*, which McNally describes as "belief in unconventional modes of causation." And they tended to be interested in New Age topics such as tarot, astrology, ghosts, and unusual forms of alternative medicine.

McNally believes his abductees' experiences all began with a phenomenon called *sleep paralysis*, combined with *hypnopompic hallucinations*—a term that refers to dreamlike visions during the transition from sleeping to waking. Both are actually quite normal, he says—no more dangerous than a case of hiccups—but they're extremely frightening and produce vivid, inexplicable memories.

Sleep paralysis normally occurs during our dreams to keep us from running around acting them out, injuring ourselves in our sleep. Usually it fades before we wake up—but not always: about one person in three sometimes awakens before the paralysis wears off. When this happens, the dream state can linger, producing vivid hallucinations we remember much better than ordinary dreams. A friend of mine calls them "waking dreams"—which, in his case, tend to happen when he's camping. He wakes up paralyzed, convinced wild animals are prowling around his tent, then, still paralyzed, fades back to sleep. In the morning he has such powerful memories of wolves, bears, or lions that he has to fight back the impulse to search for tracks.

The classic alien abduction tale has much in common with my friend's nocturnal animals: *I woke up paralyzed, felt a presence, etc.* In fact, if my friend had been an *X-Files* believer he might have put a quite different spin on his own experiences. McNally claims that abductees, with their pre-existing New Age beliefs, become understandably interested in finding out what happened to them. Often they use hypnosis or other therapies that take them back to the time of the event, perhaps with leading questions derived from the therapist's own UFO beliefs. The abductees, he argues, then "recover" the entire alien-abduction mythos, complete with sexual experimentation designed to produce a human-alien hybrid race and all the other trimmings. But instead of recovering real memories, McNally says, they've had false ones implanted—created by their own imaginations from the powerful hallucinations they had during sleep paralysis and sometimes reinforced by suggestive forms of memory-recovery therapy. None of this, he adds, is new. Other cultures simply plugged it all into their own myths, producing tales of witches and demons.[116]

Subliminal Effects

Another form of mind manipulation is subliminal

---

[116] See Kaja Perina, *Psychology Today*, March/April 2003, pp. 66-76.

advertising. The concept dates back at least to Freud and has also been kicking around science fiction for a long time, but it drew new interest in the 2000 presidential campaign, when Democrats accused Republicans of inserting the word "rats" into a political ad in an effort to program voters' subconscious minds with reflexive dislike.[117]

Subliminal advertising works by flashing pictures or text more rapidly than the conscious mind can process them. Not only are people's minds being programmed, but the victims are unaware of being manipulated. Following the 2000 presidential campaign, Joel Weinberger of Adelphi University's Derner Institute in Garden City, New York decided to test whether this actually worked by setting up a Web site on which people rated a hypothetical political candidate after being flashed with one of four subliminal messages. Those messages were "RATS," "STAR" (*rats* spelled backward), ARAB, and XXXX (a neutral control).[118] The study proved that people really don't like rats and, accordingly, the candidate subliminally associated with them. "STAR," which could be viewed as praise, had little effect, as did "ARAB" (good news for those concerned about anti-Arab bias). But RATS substantially lowered the viewers' impression of the hypothetical candidate.

In a more complex experiment conducted in the laboratory, he and colleague Lloyd Silverman had previously examined the reaction of people to the subliminal message "MOMMY AND I

---

[117] ABC News/AP, "Democrats Smell a Rat in GOP Ad," September 12, 2000.

[118] Joel Weinberger & Drew Westen, *Political Psychology*, Vol. 29, Issue 5, pp. 631-651, October 2008.

ARE ONE."[119] Prior research had indicated that if people are given subliminal flashes of this message and then asked to recall their childhoods, it often increases their tendency to remember good events. But not always. Some studies had found that the same message promoted bad memories.

Weinberger hypothesized that the subliminal message was actually received in both situations, but that the results depended on whether people had good or bad childhood experiences with their mothers. To test this, he recruited a new group of study subjects and gave them a sociological test designed to assess their relationships with their mothers. Then, after priming some with the Mommy message, he asked everyone to write down memories from their childhoods. The goal wasn't to have them write complex essays, but simply to produce lists, such as "I got a puppy for my birthday," "I went to summer camp," "We moved across town," "My sister fell off a horse and broke her collarbone," "My parents got divorced." After a few minutes, time was called, and the test subjects were asked to rate each memory as good or bad, on a multi-point scale. Those who'd revealed positive childhood experiences with their mothers on the sociological exam responded positively to subliminal stimulation with "MOMMY AND I ARE ONE." Those who'd had poor experiences in childhood didn't take comfort from that message but were instead induced to recall a higher than normal level of bad memories.

It's not an earthshaking finding, but the MOMMY AND I test does reveal some interesting things about subliminal messages. On the most superficial level, it's another

---

[119] Silverman & Weinberger, *American Psychologist*, 40, pp. 1296-1308 (1985).

confirmation that subliminal messages do indeed work. The mixed results of the earlier MOMMY AND I studies were simply due to an overlooked variable. But on a more complex level, it means that the effects of subliminal messages may vary, depending on what type of images they provoke in each recipient. Folks like Willard, the rat-loving outcast in the 1971 horror movie of the same name, would probably vote *for* a candidate about whom they'd been subliminally programmed with messages about rats. (An interesting science fictional angle on this might involve a story about a subliminal manipulation program that is uncovered when it backfires in exactly this manner, by inducing a subset of the population to do the precise opposite of what the advertisers intended.)

Kissing the Frog

However powerful they are, subliminal messages appear to be quite limited in scope. In the MOMMY AND I test, the messages affected the subjects' attitudes about their childhoods, but did not strongly alter the content of their memories. Those who'd been flashed with MOMMY AND I ARE ONE did not have larger numbers of Mother-related memories. Thus, while subliminal messaging might be used to make you *want* to go to Mars, it's not a promising method for causing you to remember a nonexistent trip.

There does not, in fact, appear to be an easy way to implant detailed memories in people who are aware of the process. But just as Leichtman has proven that it's possible to use leading questions to confuse the memories of young children, Elizabeth Loftus, a memory researcher at the University of Washington, has demonstrated that it's possible to implant some truly astounding memories in adults who are unaware they're being manipulated.[120] These memories, she found, can be extremely detailed—and while research ethics preclude tinkering with anything but relatively trivial memories, there's no reason similar techniques couldn't be used for considerably more sinister purposes.

In one study, adult volunteers were given a variety of objects, such as pink scissors or a toy frog. They were encouraged to play with these objects in whatever innovative fashions came to mind, while the researchers kept track of what they did. The following day, and on as many as four additional follow-up sessions, the subjects were asked to imagine doing additional things with these objects. These imagining sessions might contain a considerable amount of sensory detail, such as "Imagine kissing the frog. Imagine the color of the frog and the feel of it against your lips." Some of the imagined actions were familiar, such as flipping a coin. Others were truly bizarre, such as crushing a Hershey's kiss in a stapler.

Two weeks later they were asked to report what they'd actually done with the objects. Fifteen percent of those who'd gone through the full battery of five imagining sessions remembered one of the imagined activities as real—including such things as kissing the frog. That might seem like a low figure, but remember, these people had been convinced they'd kissed a toy frog or stapled a Hershey's kiss, simply by being asked to imagine what doing so might feel like. Imagination is a powerful force.

---

[120] Elizabeth F. Loftus, *Learning Memory*, 12: 361-366 (2005).

Memory Bugs

A second study applied a more subtle approach to the same end, although in this case the memories were long-ago events from childhood. Loftus recruited volunteers by telling them that she was seeking people to critique advertisements. But, like many psychology experiments, the true purpose was hidden.

Her volunteers were then shown a series of ads, including one touting the pleasures of Disneyland. The Disney ad contained a prominent picture of Bugs Bunny, along with text talking of how shaking hands with him is the type of event that can make a child's day. There was only one hitch: Bugs is a Warner Bros. character who would never be found in Disneyland. People who spotted this fallacy were labeled as "Bugs detectors" and dropped from the study. The rest were asked if they remembered meeting Bugs on their own childhood visits to Disneyland. Sixteen percent said "yes."[121]

Then the study was repeated by having a different group critique not one, but three Bugs-containing ads. Now, 36 percent remembered meeting Bugs. When pressed for details, 62 percent of those said they shook his hand, 40 percent hugged him, and at least one heard him say, "What's up doc?" Again, using a very simple ruse, Loftus and her colleagues had persuaded a startling percent of people that they'd had impossible experiences.

A third study took on the belief that memories of traumatic impacts are so strong that they're immune to such manipulation. To test this, Loftus interviewed Russians about their memories

---

[121] So did seven percent of the control subjects, whose memories had not been contaminated by the ad. Some people simply have bad memories or are so anxious to please the researcher that they'll say just about anything.

of a 1999 apartment complex bombing in which 230 people died—an event that shocked their country in much the way that the World Trade Center attacks of 2001 shocked Americans. The Russians were first interviewed in March of 2002, two-and-a-half years after the attack. Using non-leading questions, the researchers asked them to recall what they remembered of that fatal day. Six months later, the subjects were re-interviewed, but this time the interview included a leading question regarding a detail none of them had reported the first time, and which may or may not actually have happened (it wasn't in any of the news coverage of the event). Specifically, they were told: "Earlier, you mentioned a wounded animal. Do you still remember it?"

A full one-eighth of these people responded with some form of "Oh, yes." When asked to elaborate, they produced remarkably detailed stories of events that never occurred, "recalling" images such as "a bleeding cat, lying in the dust," "a lost parrot in a cage," or "a crazed dog, barking and rushing around the police officers." Again, one person in eight might not sound like much, but how many of us are sure we'll never forget all the details of what we saw on the media coverage of the World Trade Center attacks? Loftus' research indicates that some of us could easily be led to invent details—and that these inventions could, at least in general terms, be guided by a stranger who was attempting to manipulate us. To me, that's deeply disturbing.

Overall, Loftus says, there are three basic steps to implanting false memories:

1. Make the potential event plausible. It doesn't have to be something the person is predisposed to believe—as in McNally's alien abduction memories—but it does have to carry the ring of

284

potential truth. People who couldn't imagine themselves being uninhibited enough to crush a Hershey's kiss in the stapler aren't going to be easily faked into believing they did it.

2. Create a belief that the event actually happened. Each of her experiments did this in a different way, but in all of them it was a key step. In one case it was done by leading questions; in another by guided imagining sessions. In the third it was by creating a scenario (the Bugs Bunny ads) in which people's imaginations were likely to run in the desired direction without any other outside urging.

3. Embellish the belief with sensory detail.

This is not a purely theoretical exercise. In 2002, Loftus suggests that the entire population of the D.C. area were unwitting subjects in an accidental memory experiment, triggered by a report linking the Beltway snipers (who killed ten people and wounded three others in a series of 15 random shootings) to a white van. When the snipers were finally caught, it turned out that they did not own such a van. But from the moment the first van sighting was reported, white vans were seen regularly throughout the manhunt. Yes, such vans are common vehicles. But the frequency and the urgency of the sightings indicates that the expectation of seeing them became part of the terrorized area's collective imaginations.[122]

The world of *Total Recall* might not yet exist, but when it comes to mass hysteria, our collective memories really can be created wholesale.

---

[122] Elizabeth Loftus, *Nature Reviews Neuroscience*, 4, 231-234 (March 2003).

HERE BE THERE DRAGONS
Author's Note

Psychology, at its core, focuses on the fundamental question of what it is that makes us human. But two more fundamental questions are what it means to be alive, and whether we are unique in the Universe. The first is the realm of philosophers and theologians. The latter is the realm of astrobiologists.

Skeptics have called astrobiology a science without a subject, meaning you can't really study alien life until you discover it. But it's possible that we may not have to go out of the Solar System to find it. And whatever many people might think, Mars may not be the best place to look.

SATURN'S "JET-PROPELLED" MOON AND THE SEARCH FOR EXTRATERRESTRIAL LIFE

It was a shame, Aimee Miller thought, that the first alien life form discovered by humanity had to be killed in the process of looking at it. But such was the way with electron microscopy: in order to see the alien cell in enough detail, it was necessary to hit it with radiation guaranteed to kill it. Just as, afterward, it was necessary to chemically pick it apart to prove that this tiny bit of one-time life melted out of an ice grain wasn't something that had somehow hitchhiked from Earth.

But if there was one microbe, hopefully there were more. Aimee wasn't sure exactly

287

where in its elaborate gravitational dance with Saturn's moons the Saturn orbiter currently was, but one thing was clear. The tiny grain of ice it had already plucked from Saturn's tenuous E ring would forever change humanity's view of its place in the Universe. And as the mission progressed, the orbiter would be making repeated flybys of the mother lode of enigmatic ice grains: the tiny moon Enceladus, where for millions of years ice geysers had spewed microscopic snowflakes from underground cauldrons now proven to be cradles of life beyond Earth. Simply stated, this small, frozen worldlet had just become the hottest place in the Solar System.

Far fetched? Not as much as you might think. The most exotic part might be Aimee's probe's fancy bio-lab, which can do complex analyses no space mission has yet dreamed of. As for seeking life at Saturn? Many scientists think that might actually be the best (and easiest) place to look. Sure, NASA's Curiosity rover has repeatedly confirmed what many have suspected: that once upon a time, Mars was habitable. But that was billions of years ago. If you switch the question to "Where, right now, might we find conditions suitable for life?" Mars becomes less interesting. In fact, the best options might be in places that barely crossed astrobiologists' radar as little as a few decades ago.

In Larry Niven's classic Known Space series of science fiction stories, the first interstellar colonies were scouted by robotic probes designed to search out habitable worlds. Due to a programming error, however, what the probes reported as habitable planets turned out simply to be habitable "places"—small niches on otherwise bleak planets. In Niven's stories, this was something of a running joke as unsuspecting colonists were forced to adapt to worlds a lot less Earthlike than anticipated. But Niven's little gag may have been unexpectedly visionary. When it comes to seeking potential abodes for extraterrestrial life, the approach many scientists have been taking—with its focus on Mars-roving and exoplanet-hunting—might be unnecessarily terra-centric (or at a minimum, planet-centric). There might, in fact, be many more-exotic places where life might exist.

One of the first to be recognized was Jupiter's moon Europa, which appears to have a planet-girdling sea beneath its icy surface. But for the moment, the astrobiological spotlight is on a much more accessible world: Enceladus—a 300-mile-wide moon of Saturn that many think might be the Solar System's best place to look for extraterrestrial life. "It has liquid water, organic carbon, nitrogen [in the form of ammonia], and an energy source," says Chris McKay, an astrobiologist at NASA's Ames Research Center at Moffett Field, California. "There is no other environment in the Solar System [except Earth] where we can make all those claims."

Superficially, Enceladus seems a singularly uninteresting place to visit. White, round, and icy, it looks from a distance like nothing more than an overgrown cue ball. Named for a giant from Greek mythology, it isn't even all that big; Saturn alone has five larger moons. Enceladus could snuggle nicely between Baltimore and Boston with room to spare. But sometimes, nomenclature is irony. In Greek lore, the giant Enceladus was buried under Italy's Mt. Etna, one of the world's most active volcanoes. In some versions of the story, in fact, it's his sleeping breath that fuels Etna's fires. And while nobody could have foreseen it when British astronomer William Herschel first spotted Enceladus in 1789, it turns out that the not-so-giant moon behaves in some ways like an enormous volcano.

"It's the only unambiguously cryovolcanically active moon in the Solar System," Dennis Matson, a planetary scientist at NASA's Jet Propulsion Laboratory (JPL) in Pasadena, California, said at a May 2011 gathering of Enceladus experts at the SETI Institute, in Mountain View, California.

As the term indicates, cryovolcanoes don't spew lava—at least not lava as we know it. Rather, they extrude a mix of water and ammonia that's only hot compared to everything else nearby. And in Enceladus's case it's not really lava, but geyser-like jets of ice spicules and chilly vapor—not so much an extraterrestrial Etna as a deep-frozen Yellowstone.

Although some people will argue that this isn't really a volcano, what makes Enceladus's cryovolcanoes, cryo-geysers, cryo-fumeroles, or whatever you want to call them interesting is that their icy vapor most likely originated as liquid water. If so, somewhere beneath the surface there may be an environment warm enough for life as we know it. Better yet, in Enceladus's low surface gravity (slightly more than 1.1 percent that of Earth's) many of those ice grains are blasted all the way into space, where a properly equipped spacecraft might be able to swoop close and collect samples, including, perhaps, a few stray microbes caught up by the jets. To look for life signs on Mars, we have to land rovers and drill into rocks for samples. On

Enceladus, we might simply have to fly by and let the samples come to us.

Tiger Stripes

The first inkling Enceladus might be more than simply an iceball came in 1980 and 1981 when NASA's two Voyager spacecraft sped through the Saturn system, passing within 120,000 miles of Enceladus en route to Uranus, Neptune, and the edge of the Solar System. Even from that close, the cameras of the day weren't good enough to reveal more than tantalizing hints, but one of those was that the surface, while not truly cue-ball smooth, was unusually free of big craters. It was a sign that something had "resurfaced" it relatively recently, erasing traces of old impacts. Also, the Voyagers found that Enceladus orbits in the heart of Saturn's tenuous E ring (so faint it wasn't even seen until 1967), raising the possibility that, somehow, the ring and the moon were linked—most likely with the ring being created by material somehow being ejected from Enceladus's surface. But that was pretty much all we knew until NASA's Cassini spacecraft arrived on the scene in 2004, more than two decades later.

Initially, Cassini's prime target was Saturn's truly giant moon, Titan, which has revealed a remarkably complex surface topography and a weather cycle in which methane plays the role water plays on Earth, complete with lakes, seas, streambeds, thunderstorms, and seasons. But it didn't take scientists long to look at Enceladus as well, on the first of some two dozen flybys scheduled before the mission ends in 2017.

What they found became more exciting with each flyby.

One of the first discoveries was of strange, bluish bands near the south pole, quickly dubbed "tiger stripes." There were four major ones, now named for cities in *The Arabian Nights*—Alexandria, Cairo, Baghdad, and Damascus—lying in parallel, roughly 130 kilometers long and 35 kilometers apart. Clearly fault lines of some sort, they confirmed the Voyager-era discovery that Enceladus wasn't simply frozen in time, but bears the unmistakable scars of recent geological activity.

But tectonic features are a dime a dozen in the outer solar system. The tiger stripes were interesting, but not game-changing.

The next surprise came a couple of months later, when a subsequent flyby revealed that the stripes weren't just cracks in Enceladus's icy façade—they were hot, at least by the standards of an airless moon, 870 million miles from the Sun, where the average temperature is somewhere around -330°F. In places, the tiger stripes have now been measured with temperatures as high as -139°F—still far colder than anything ever measured on the surface of Earth, but a strong sign of an internal heat source. It was a bit like going to Antarctica and finding places hotter than anything ever recorded in the Sahara. Clearly, something interesting was going on.

But the biggest find came later that year, when Cassini scientists released a photo showing a cloud of backlit particles jetting from Enceladus's south polar regions, right above the tiger stripes. It would have made a great cover for a science fiction novel, looking almost as though the moon was rocket propelled, blasting northward out of its orbit. That's not actually the case, of course. (Nor is there enough impulse from the plume to overcome other forces holding Enceladus in its orbit.) Rather, the plume is Matson's cryovolcano: the jets from what are now known to be dozens of geysers spewing a total of 440 pounds of water vapor and ice particles into space each second—enough to fill an Olympic-sized swimming pool every few hours. It was evidence that Enceladus is indeed the source of the E ring. But the photo was so startling that the scientists took months before releasing it, double-checking their instruments to make sure the plume was real and not some odd type of lens flare or other instrumental artifact.

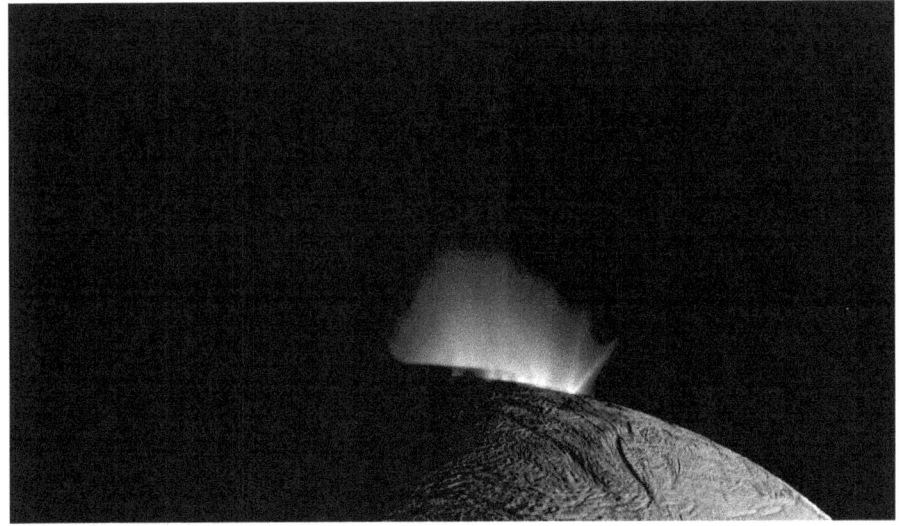

Cassini Imaging Team, SSI, JPL, ESA, NASA

Since 2005, Cassini has already made nearly 20 more flybys, one dipping as close as 13 miles to the surface, sampling the plume by flying right through its heart. Other flybys have photographed the surrounding terrain, made increasingly detailed thermal maps, and used spectroscopes to analyze the composition, speed, and density of the plume's dust grains and gas. One thing that's now known, for example, is that the stripes are valleys, each about 1,600 feet deep and 1.2 miles wide. Flanking them on each side are low ridges, about 300 feet high, giving the whole thing a vaguely "M" shaped-profile. (Although in this "M" the dip in the middle would extend well into the next line of type.) Even on Earth, these would be dramatic features, roughly comparable to Colorado's Black Canyon of the Gunnison. On much smaller Enceladus, they are enormous.

Not surprisingly, the geyser-like jets feeding the plume originate from vents in the tiger stripes, corresponding nicely to the hot spots. But this doesn't mean the tiger stripes are jetting vapor everywhere. Rather, the individual jets appear to be quite small, by some calculations (based on exhaust velocities of the particles) as narrow as a few tens of centimeters—roughly a foot.

Between the jets and the heat, Enceladus appears to be emitting an enormous amount of energy into space: about 16 gigawatts, Enceladus scientists have calculated—enough to power Washington, D.C., and several of its suburbs. And with all of that energy comes not only water vapor and ice crystals, but a stew of other intriguing chemicals: ammonia, methane, carbon dioxide, hydrogen cyanide, formaldehyde, acetylene, and other hydrocarbons. None of this means that Enceladus has life. But with each successive find, the possibility of a habitable zone not far beneath its surface becomes ever more likely. Water, carbon, nitrogen, and an energy source…Cassini wasn't sent to Saturn looking for life. But what it's found with instruments designed for other purposes makes one wonder what a future mission, with a more specialized instrument package, might someday uncover.

The big question, of course, is what is producing the plumes. Most of the energy needed to drive them probably comes from gravitational flexing, as Enceladus moves though Saturn's enormous gravity field. That causes the moon's interior materials to rub against each other, generating heat. But even if you also throw in heat from radioactive decay, there doesn't seem to be enough for 16 gigawatts of energy—which means there's either a missing energy source or the plume isn't active all the time. In fact, says Frank Postberg, a physicist at Heidelberg University in Germany, "the energy output suggests that [the plume] cannot be active much more than 10 percent of the time, at least at the current scale."

Many Enceladus scientists therefore believe that instead of blasting away continuously, the plumes probably follow what engineers call a duty cycle, in which they turn on and off at intervals, accumulating energy for future eruptions during the inactive periods, much as earthly geysers do, but on a longer time scale. Nobody knows how long the "on" cycles might last. All we know for sure is that the plumes have been going at full blast ever since Cassini first saw them. And since they do feed the E ring, it's a pretty good guess they've been doing so at least since the E ring was first spotted, 47 years ago. In fact, they've presumably been active long enough for the E ring as we know it to have formed—a process Postberg estimates would take somewhere between a few hundred and "maybe" a thousand years.

But maybe there's another marker for how long they've been active, says Paul Schenk, a planetary scientist at the Lunar and Planetary Institute in Houston, Texas.

Snow.

HERE BE THERE DRAGONS
Talcum Powder

When the plumes are operating, Enceladus isn't just a cold, airless world. It has weather. Cold and sunny, with a chance of snow flurries. That's because many of the ice grains ejected from the geysers don't get far into the E-ring before Enceladus reclaims them. "The vast majority, we think, fall back to the surface," says Carolyn Porco, head of the Cassini imaging team.

But the snowfall isn't evenly distributed. In a 2010 study in the journal *Icarus*, a team led by Sascha Kempf, then of the Max Planck Institute in Germany, calculated that the dual effects of Enceladus's and Saturn's gravity would concentrate it into two narrow bands, stretching hundreds of kilometers northward from the tiger-stripe zone.[123]

Based on this, Schenk went looking for snowdrifts. Using the highest-resolution photos available for the fallout zones, he found a 20x20 kilometer region where successive flybys had taken photos from different angles, allowing him to construct 3D "stereo" images showing an area whose topography was muted into rounded contours indicative of deep snow, with the ghostly outlines of underlying features peaking through. "It's different from what you see in other areas," he says. "This is much smoother. It really does look as if it's been mantled by a deposit."

The superfine ice crystals (finer-grained even than talcum powder, Schenk says) might make for the Solar System's best powder skiing. But the goal wasn't to find an exotic science-fictional vacationland. It was to measure the depth of the snow—something Schenk was able to do by using stereo images

---

[123] Sascha Kempf, et al, *Icarus*, Volume 206, Issue 2, April 2010, pp. 446–457.

of snow-draped canyons, up to 1,600 feet deep and a mile across.

From these photos, Schenk was able to distinguish the overlying snowpack from the underlying topography. "Slope breaks near ridge crests can be interpreted as a weak layer [snow] over a solid layer [underlying ice]," he says. "I get a [snow] thickness of 125 meters—give or take 50 meters. It varies from site to site."

That's a lot of snow, anywhere. But on Enceladus, it's enormous because Kempf's model doesn't show it coming down at blizzard rates. Rather, his team calculated that even in the highest snowfall zones it was accumulating at only a rate of a millimeter every 1,500 years. At that rate, it would take nearly 200 million years to pile up 125 meters. That said, there's a bit of apples and oranges going on in these calculations, because Kempf's calculations were based on highly compressed snow, and Schenk's drifts might be quite fluffy—meaning they could pile up a lot deeper, faster. But even if there's a factor of ten difference, Schenk's finding still suggests that the plume must have been active for nearly 20 million years. And even in Enceladus's low gravity, snow at the bottom of that giant pile would be substantially compacted, which means 20 million years is a lower-bound estimate.

But the point isn't to debate whether the plume has been going for 20 million years or 200 million. The point is that the plumes aren't a brand-new feature. "It's a big step from a thousand to 10 million years," says Postberg. "This is a whole new order of constraints for how long the plumes have been active." Furthermore, it's possible that Postberg's duty cycle doesn't require the plumes to shut off entirely. Rather, they may ramp up and down, without ever totally subsiding. If so, the average rate of snowfall on the rim of Schenk's canyon might be much lower than today, and Schenk's snowfall record might extend not merely over tens of millions of years, but over hundreds of millions. And that's important because the longer the geysers have been active, the greater the chances that the processes producing them might have given life enough time to evolve.

Urine Dumps

Life as we know it needs liquid water, not snow. So the mere fact the tiger stripes are jetting out moisture doesn't necessarily mean there's liquid beneath. The jets could be fueled by "dry" processes in which ice grains and vapor are spewed out without melting.

One such process could involve sublimation (in which ice evaporates directly into vapor, without ever getting warm enough to melt). Another might involve the decomposition of subsurface layers of clathrate deposits (a type of ice composed of a cage-like lattice, tightly woven enough to trap gas molecules). Both processes occur on Earth. Fishermen often dredge up methane-containing clathrates (also called methane hydrates) from the Arctic waters, watching them decompose into water and methane. And villagers in Greenland use sublimation to air-dry laundry on cold, sunny days when temperatures never get above freezing.

But these processes produce mostly water vapor, not the 50/50 mix of vapor and ice crystals observed by Cassini's instruments, says Andrew Ingersoll, a planetary scientist from California Institute of Technology in Pasadena. "It's hard to get solid-to-gas ratios of more than one percent if you think all the particles are forming from vapor," he says. Water droplets, on the other hand, easily form clouds of tiny ice particles. "If you throw a blob of liquid into [a] vacuum," he says, citing astronaut "urine dumps" as an example, "it's not going to freeze into a big chunk. It'll break up and explode into a cloud of smaller particles."

An even worse blow to the "dry" theories comes from what happens when the ice grains are hit by Cassini's cosmic dust analyzer as the spacecraft flies through the heart of the plume. These collisions are energetic enough to produce a quick flash of light as the impact not only vaporizes the ice grain but ionizes its constituents, allowing scientists to determine the grain's composition from its spectra. In a 2001 paper in *Nature*, Postberg's team saw clear indications in these grains of sodium chloride: salt.[124]

Salty ice particles, he says, are hard to explain with "dry" plume models, because even if salt is present underground, sublimation or clathrate degeneration should either leave it behind or bring up sodium vapor in the plume, something that has not been seen. For appreciable quantities of salt to be in the ice grains, he says, they must have begun as frothy bubbles fizzing out of a salty, subsurface ocean, rich in dissolved gases: bubbles that then collect into a wind that blows up through narrow fissures, emerging at the surface as the geyser-like jets seen from space.[125] Matson calls this a "Perrier ocean" and compares it to being on a beach on a blustery day. "When the

---

[124] F. Postberg, et al, *Nature* 474, 620–622 (30 June 2011).

[125] The process that produces this is similar to that seen in Mars's black spiders (see chapter on Mars), although the causes are different.

surf is up," he says, "there's a lot of spray. You can smell the salt in the air." The amount of salt, Postberg adds, is about half that in Earth's oceans—comparable to what you'd get if you put liquid water in contact with primordial rock for millions of years.

So far, we've been talking as though the existence of an underground ocean is all but proven. But nothing on the frontiers of science is that cut and dry, and not all the evidence points to liquid water.

One problem is the energy problem, which doesn't give enough heat to maintain a layer of water beneath the ice, everywhere, says Francis Nimmo, a planetary scientist at the University of California, Santa Cruz. But that's just a minor difficulty, because if all of the gravitational energy is focused in a single zone beneath the tiger stripes there is plenty to produce a large, subsurface sea. And once such a sea forms, Nimmo says, it would concentrate gravitational flexion (and therefore heating) into its own region, thereby becoming self-perpetuating. In fact, he says, it's possible that such a regional sea might be a remnant of a once-larger ocean that long ago mostly froze. Another prospect is that such a sea was created by a long-ago asteroid impact. It's a theory that draws support from the fact that the southern part of Enceladus, which contains the tiger stripes, is separated from the rest of the moon by a "dichotomy boundary" marked by features that might (or might not) be remnants of exactly such a crater rim.

On the other hand, complex fracture patterns in the ice around the tiger-stripes region (including ones that appear to be remnants of several generations of previous tiger stripes) indicate that the entire surface of that part of Enceladus may have slowly rotated due to asymmetric torques from Enceladus's not-perfectly-circular orbit. There's no way to do that without a global ocean under the ice, says Simon Kattenhorn, a planetary geologist at the University of Idaho. "You can't have just the south-polar terrain rotating above a local sea."

Another problem comes from the zoo of compounds found within the plume. Some, says Hunter Waite, a space physicist at the Southwest Research Institute in San Antonio, Texas, are inconsistent with a liquid source because they are ones like hydrogen cyanide, which, if they had ever met liquid water, should have reacted to produce other compounds not yet found. Others, like methane, simply aren't soluble in water.

In discussing this, however, Waite is careful not to argue there's no liquid water. He merely notes that some of these chemicals are consistent with liquid water…and some aren't. "We don't know how to sort that out," he says. It's possible, he notes, that chemicals in the plumes might come from multiple sources, some involving water, some not.

This isn't a debate that's likely to be resolved in the near future. In the latest pro-ocean salvo, a 2013 study in *Nature* spearheaded by Matt Hedman, a planetary scientist at Cornell University, reported that the intensity of the jets feeding Enceladus's plume varies with Enceladus's distance from Saturn.[126] "The jets of Enceladus apparently work like adjustable garden hose nozzles," Hedman said in a NASA press release. "The nozzles are almost closed when Enceladus is closer to Saturn and are most open when the moon is farthest away. We think this has to do with how Saturn squeezes and releases the

[126] M. M. Hedman, et al, *Nature*, 8 August 2013, pp. 182–184.

moon with its gravity."[127] That itself might simply be an intriguing factoid. But Hedman's coauthor, JPL planetary scientist Christophe Sotin, argues that it wouldn't work that way under "dry" models of plume formation. "The way the jets react so responsively to changing stresses on Enceladus suggests they have their origins in a large body of liquid water," he said in the same press release.

^{12}C, ^{13}C, and Acetylene

Even if Enceladus has liquid water, that's not proof of life. Just to begin with, there has to be some way for microbes to turn the available energy into something they can use. One option obviously not available in Enceladus's dark underground is photosynthesis, the process by which earthly plants (and some microorganisms) harvest energy from sunlight. But a process that might be available, says McKay, is a "methanogen system" in which microorganisms live by obtaining both energy and biological building blocks by synthesizing methane from carbon dioxide.

McKay is particularly fond of the idea because it might be sustainable over very long time periods if geological processes carry some of the methane produced by the bacteria back into zones hotter than 500°C. Those temperatures would decompose it back into building blocks usable by methanogen bacteria, meaning that overall, Enceladus could have an ecosystem in which organisms subsist on chemicals recycled by geochemical heat. "The fact we see methane in the plume makes it interesting," he says.

---

[127] http://saturn.jpl.nasa.gov/news/newsreleases/newsrelease20130731/.

Ronald Oremland, a microbial biogeochemist with the U.S. Geological Survey's office in Menlo Park, California, believes an even better food source would be acetylene. On Earth, this chemical doesn't occur naturally, though humans synthesize it as fuel for welding torches. But it exists in comets, and there are hints it might also be present in Enceladus's jets. And acetylene-eating organisms do exist on Earth. If there's enough acetylene on Enceladus, Oremland says, it could be "fast food" for microbes—a primordial food source on which Enceladus bugs might still be munching away. If so, possible biomarkers for such organisms might be byproducts of acetylene metabolism, such as acetate and acetaldehyde.

Other possible biomarkers are amino acids, especially if they can be tested for chirality, which is the degree to which they mix mirror-imaged shapes known as D and L isomers. Abiotic processes tend to produce an even mix of the two isomers. Biological ones favor one or the other. On Earth that's the L version, though there's no known reason extraterrestrial life couldn't do the opposite. "If we find amino acids and there's a strong chiral preference, that's persuasive evidence for a biological origin," McKay says.

But the best biomarkers, McKay and Oremland agree, might be found in the ratios of carbon's two stable isotopes, ^{12}C and ^{13}C. That's because biological processes produce compounds slightly enriched in ^{12}C, compared to nonbiological ones. (Testing $^{12}C/^{13}C$ ratios, in fact, is one way sports authorities can catch drug-cheating athletes, because synthetic hormones, produced in a lab, have different isotope ratios from those

303

produced by the athletes' own bodies.)[128] Thus, if the methane coming from Enceladus is unexpectedly rich in ^{12}C, it's a strong indication it's being made by bacteria. Similarly, if the plume is emitting acetylene byproducts that are richer in ^{12}C than Enceladus's primordial acetylene, it suggests that acetylene-eating bacteria may be at work.

Unlike the electron microscope on Aimee's hypothetical spacecraft, the equipment to conduct carbon-isotope tests isn't all that elaborate. It just happens not to have been included in Cassini's payload…which means that carrying out such tests would require a return to Enceladus.

Several such missions are possible, said Nathan Strange, a mission architect at NASA's Jet Propulsion Agency in Pasadena, California, at the 2011 SETI Institute meeting. The simplest would merely be to go back to Saturn with a dedicated Enceladus mission. But it might be more efficient to use a multi-purpose mission that piggybacked an Enceladus orbiter onto another dream project: a Titan rover. A fancier mission might be an Enceladus lander, or possibly a "hopper" capable of launching itself from one landing zone to another. "The low gravity makes this possible," Strange said. "You could even hop across a jet."

A sample-return mission is also possible, in which a probe flies through Enceladus's plume, collects some ice-grains, and hurls them back to Earth for analysis. "This is really the low-hanging fruit" of sample-return missions, Peter Tsou of Sample Exploration Systems in La Canada, California, told space.com in

---

[128] This is also important in distinguishing geochemical processes on Earth from biogeochemical ones. See chapter in Snowball Earth.

2012. "It would be a shame not to pick it."[129]

Not that any of this can occur instantly. Every couple of decades it's possible to use Jupiter as a gravity boost en route to Saturn, but the nearest window for that is most likely lost. "We're probably looking at 9 to 10 year flight times," Strange said.

Still, Enceladus presents a unique opportunity, both for scientists seeking answers to the ultimate question and science fiction fans who enjoy playing mental games of "what if?" "What we're being handed at Enceladus is a potential gift of looking at life in the outer Solar System," says Oremland. "What's appealing about Enceladus is that you have some of the conditions for life. There's liquid water under the ice. It seems to have been around a long time. How long, nobody knows. One hundred million years? A billion? That's a long time for life to get going, provided there's something to eat."

And for science fiction fans, Enceladus provides a world unlike anything the best minds in the field had ever, on their own, imagined.

From *Dune* to Dyson spheres, from *Game of Thrones* to "Nightfall," one of the mainstays of science fiction is "worldbuilding"—the imagining of scientifically plausible planets sufficiently unlike our own to be worth a story...but not so implausible as to fall into the realm of fantasy. Sometimes, these are extrapolated from contemporary Earth. Sometimes, they involve places like Arrakis, Trantor, or the Ringworld, with no known real-life analogs. But sometimes, they are inspired by worlds like Enceladus, where reality outstrips imagination and science and wonder collide.

---

[129] Mike Wall, "Saturn Moon Enceladus Eyed for Sample-Return Mission," *space.com*, December 6, 2012.

ABOUT THE AUTHOR

Richard A. Lovett grew up never quite sure what he wanted to do. He loved science, history, law, and the outdoors. For years, he wanted to be a cosmologist—enough so that in college he majored in astrophysics. When trying to decide what to do in grad school, he wrote to 27 programs: nine in cosmology, nine in law, and nine in economics. Offered a position in the University of Michigan's program in law and economics, he jumped at the  chance to get two of three in a single package.

Later, he taught law at the University of Minnesota, worked for an environmental consulting firm, taught environmental studies at California State University, Sacramento, and researched water law at Oregon State University.

In the interim, he took a summer off to bicycle solo across the country. He wrote a book about it and soon was writing full-time, though even by journalistic standards his interests were diverse: environmental policy, travel, sports, legal news, and science—more than 3,500 articles, so far.

In 1999, Stanley Schmidt, then editor of *Analog Science Fiction & Fact* magazine (in which Isaac Asimov got his start), invited him to submit a science article. Since then, he's appeared in *Analog* more than 150 times, including more than 40 works of fiction. Other fiction has appeared in *Nature*, *Cosmos*, *Running Times*, *Marathon & Beyond*, *Abyss & Apex*, and has been translated into Russian, Portuguese, Finnish, and Polish.

It's a combination of interests that makes him unusually drawn to topics like those in this book: findings on the frontiers of science, where the social ramifications are often as fascinating as the science itself.